CATALYZED CONTROLLED CRYSTALLIZATION OF GLASSES
IN THE LITHIUM ALUMINOSILICATE SYSTEM

KATALIZIROVANNAYA REGULIRUEMAYA KRISTALLIZATSIYA STEKOL
LITIEVOALYUMOSILIKATNOI SISTEMY

КАТАЛИЗИРОВАННАЯ РЕГУЛИРУЕМАЯ КРИСТАЛЛИЗАЦИЯ СТЕКОЛ
ЛИТИЕВОАЛЮМОСИЛИКАТНОЙ СИСТЕМЫ

Part 1

CATALYZED CONTROLLED CRYSTALLIZATION of GLASSES
in the
Lithium Aluminosilicate System

Edited by
V. V. Vargin

Authorized translation from the Russian by
E. B. Uvarov, B.Sc., A.R.C.S., D.I.C., A.R.I.C.

SPRINGER SCIENCE+
BUSINESS MEDIA, LLC
1965

The Russian text, published by Khimiya Press in Leningrad in 1964,
was corrected and revised by the editor.

Library of Congress Catalog Card Number 65-25265

© 1965 Springer Science+Business Media New York
Originally published by Consultants Bureau Enterprises, Inc. in 1965
Softcover reprint of the hardcover 1st edition 1965

ISBN 978-1-4899-4910-3   ISBN 978-1-4899-4908-0 (eBook)
DOI 10.1007/978-1-4899-4908-0

Preface

During the past four or five years, new crystalline glass materials with a number of valuable physicochemical properties have attained special importance among the new glassy materials. These glass-ceramics are known as sitalls in the Soviet literature and as Pyrocerams (U.S.A.) or vitrokerams (Germany) in the foreign literature.

Glass-ceramics can be produced from a whole range of vitreous materials and may have the widest applications in industry and science in accordance with the properties obtained. It is impossible at the present time to foresee all the potentialities of glass-ceramics; it is therefore necessary to study, both theoretically and practically, all aspects of the processes occurring during their formation and the properties of these materials.

Glass-ceramics are formed as the result of controlled fine-grained volume crystallization of glass during heat treatment.

Controlled volume crystallization is a new problem in investigation of processes occurring in glass. This problem is becoming of primary importance and the number of investigations on the subject has increased continuously during recent years.

Numerous physical and chemical methods of investigation must be used for study of all aspects of the controlled crystallization process. In order to obtain as unambiguous a picture as possible of the course of volume crystallization during heat treatment and to make complete correlation of the results possible, the authors decided to conduct the investigation with glass of one composition, subjecting it to different heat treatments.

The use of a variety of physical and chemical methods required the participation of a large number of workers: A. G. Alekseev (X-ray diffraction), V. V. Vargin (the leaching and color indicator methods, and optical density determinations), V. N. Vertsner and G. P. Tikhomirov (electron microscopy), N. E. Kind (optical crystallography and density determinations), Yu. N. Kondrat'ev (investigation of electrical properties and compilation of the survey in Chapter III), E. V. Podushko (the polythermal method), N. A. Tudorovskaya (differential thermal analysis, and determinations of refractive indices and linear thermal expansion), and V. A. Florinskaya (infrared spectroscopy). In addition, M. V. Zasolotskaya, E. M. Milyukov, V. I. Skorospelova, É. F. Cherneva, and K. A. Yakovleva took part in the experimental work. This multiple investigation was undertaken in order to obtain the fullest possible information on the catalyzed crystallization process and to elucidate the potentialities of each of the methods separately and in conjunction with each other.

The general survey of the literature on the theory of the crystallization process (Chapter II) was compiled by M. V. Serebryakova. The survey covers mainly recent publications, but some individual earlier papers are cited where necessary. In addition, a list of the literature used is given at the end of each chapter, and at the end of the book there is a general list of publications on catalyzed crystallization (beginning with 1960) not included in the bibliographies appended to the individual chapters. A bibliography of earlier publications is given in A. I. Berezhnoi's book ''Photosensitive Glasses and Crystalline Glass Materials of the 'Pyroceram' Type.''

N. E. Kind and N. A. Tudorovskaya were responsible for the literary presentation of the material. The general editor was V. V. Vargin.

This research was the first part of an investigation of crystallization of a series of glasses in the $Li_2O - Al_2O_3 - SiO_2$ system, catalyzed by titanium dioxide.

The authors are deeply grateful to Academician A. A. Lebedev for his constant interest and for valuable advice and comments.

Contents

Publisher's Note

The following Soviet journals cited in this book are available in cover-to-cover translation:

Russian Title	English Title	Publisher
Doklady Akademii Nauk SSSR	Soviet Physics-Doklady	American Institute of Physics
Izvestiya Akademii Nauk SSSR: Otdelenie khimicheskikh nauk	Bulletin of the Academy of Sciences of the USSR: Division of Chemical Science	Consultants Bureau
Kolloidnyi zhurnal	Colloid Journal	Consultants Bureau
Optika i spektroskopiya	Optics and Spectroscopy	American Institute of Physics
Steklo i keramika	Glass and Ceramics	Consultants Bureau
Zhurnal fizicheskoi khimii	Russian Journal of Physical Chemistry	Chemical Society (London)
Zhurnal neorganicheskoi khimii	Journal of Inorganic Chemistry	Chemical Society (London)
Zhurnal prikladnoi khimii	Journal of Applied Chemistry USSR	Consultants Bureau
Zhurnal strukturnoi khimii	Journal of Structural Chemistry	Consultants Bureau

CHAPTER I

Introduction

Crystalline glass materials, known as glass-ceramics, or as sitalls in the USSR, are made by the usual methods of glass technology—by melting glass batches containing crystallization catalysts in pot or tank furnaces. The articles are formed by blowing, pressing, rolling, or casting and are then subjected to another heat treatment during which a considerable proportion (20-99%) of the glass is converted into the crystalline state [1,2].

Ordinary crystallization begins at the surface (surface crystallization) and extends into the material. It usually occurs at temperatures above the softening temperature of the glass. Some individual crystals may grow to a large size.

Catalyzed (volume, controlled, directed) crystallization occurs uniformly throughout the glass and at lower temperatures than surface crystallization. In this case, the crystal size ranges from 100 A to 1 μ, with 10^9-10^{12} crystals per mm^3. Such fine-grained and uniform crystallization determines the very good mechanical and thermal properties of crystalline glass materials. Other properties of these materials depend on the chemical composition of the original glass and on the composition of the crystals formed [1-3].

Catalyzed crystallization of glasses is being widely studied both in the USSR [1,4] and abroad [5], but the processes involved are far from being fully understood, and a complete and consistent theory of catalyzed crystallization is still lacking.

Controlled volume crystallization in glass depends on the formation of numerous crystallization centers or nuclei distributed uniformly in the glass; these may be submicroscopic crystals of the catalyst or, if immiscibility effects occur in the glass, microseparation regions. In the latter case, crystallization begins at the interphase boundaries. Because of the extremely large numbers of nuclei, the crystals of the main crystalline phase cannot grow to a large size and remain in the range of 1-2 μ or less.

The present investigation was concerned with catalyzed crystallization of one glass in the system $Li_2O - Al_2O_3 - SiO_2$, containing titanium dioxide as the crystallization catalyst and having a tendency to phase separation. Glasses of this system are of great practical importance, since they, in addition to glasses of the systems $MgO - Al_2O_3 - SiO_2$ [6] and $Li_2O - Ga_2O - SiO_2$ [7], can yield transparent crystalline glass materials. A particularly important point is that only the system $Li_2O - Al_2O_3 - SiO_2$ gives rise to transparent crystalline glass materials with zero or small negative or positive values of the coefficient of thermal expansion [8,9].

The phase diagram of the system $Li_2O - Al_2O_3 - SiO_2$ is very complex and has not been fully studied, although there have been numerous investigations of this system and of certain of its sections (see Chapter III).

The material used in our investigation was a glass close to spodumene in composition; addition of titanium dioxide as the crystallization catalyst and heat treatment over a wide range of temperatures made it possible to obtain transparent and opaque crystallized specimens.

The composition of this glass was chosen on the basis of an earlier investigation, with the aid of a variety of methods, of a number of glasses in the system $Li_2O - Al_2O_3 - SiO_2$ containing TiO_2 as catalyst and forming transparent crystalline glass materials.*

---

*That investigation was carried out by E. V. Podushko. He also performed the heat treatment of specimens of the glass under investigation and prepared glasses with various contents of TiO_2.

LITERATURE CITED

1. A. I. Berezhnoi. Photosensitive Glasses and Crystalline Glass Materials of the "Pyroceram" Type, VINITI (1960).
2. S. D. Stookey. Glastech. Ber. 32(5):1–8 (1959).
3. I. I. Kitaigorodskii. Glass and Glass Making, Promstroiizdat (1950), p. 32.
4. The Glassy State, Part 1. Catalyzed Crystallization of Glass, Izd. Akad. Nauk SSSR (1963). [English translation: The Structure of Glass, Vol. 3, Consultants Bureau, New York (1964).]
5. 63rd Annual Meeting of the American Ceramic Society, Toronto, 1961.
6. M. I. Kalinin and E. V. Podushko. in: The Glassy State, Part 1, Izd. Akad. Nauk SSSR (1963), p. 164. [English translation: The Structure of Glass, Vol. 3, Consultants Bureau, New York (1964) p. 175.]
7. G. T. Petrovskii, E. N. Krestnikova, and N. I. Grebenshchikova. in: The Glassy State, Part 1, Izd. Akad. Nauk SSSR (1963), p. 167. [English translation: The Structure of Glass, Vol. 3, Consultants Bureau, New York (1964) p. 177.]
8. I. M. Buzhinskii, E. N. Sabaeva, and A. N. Khomaykov. in: The Glassy State, Part 1, Izd. Akad. Nauk SSSR (1963), p. 127. [English translation: The Structure of Glass, Vol. 3, Consultants Bureau, New York (1964) p. 133.]
9. E. J. Smoke, J. Am. Ceram. Soc. 34(3)(1951).

CHAPTER II

Survey of the Literature on Crystallization Processes in Glasses

Problems of what is known as controlled volume crystallization of glasses form the subject of a large number of recent publications in which various aspects of crystallization are examined. There is as yet no single unified theory of glass crystallization since the processes involved are extremely complex and are influenced by numerous factors. A brief account of recent work in this field is given below.

Crystallization is undesirable in the technological manufacture of ordinary glass products. The causes of this effect, the optimum conditions for preventing crystallization during melting, shaping, and subsequent treatment of glass, or for reducing it to a minimum, are examined in many publications, and in particular detail by Mukhin and Gutkina [1].

Special interest in controlled volume crystallization of glass is associated with the development of new crystalline glass materials. These materials, originally obtained from photosensitive glasses and having many valuable properties, have become the subject of independent investigations. A detailed review of published work on photosensitive glasses and crystalline glass materials has been made by Berezhnoi [2]. In this review he examines the theory of spontaneous formation of crystallization nuclei (centers) and crystal growth in relation to photosensitive glasses containing small additions of metallic salts (silver, gold, platinum, copper).

In the present survey the main attention is focused on processes in glasses without photosensitive additives. The properties and practical applications of glass-ceramics are not examined.

The modern theories of formation of crystal nuclei are based on thermodynamic principles first put forward by Gibbs and subsequently developed by Volmer. These general considerations apply to any solutions or melts. From the thermodynamic standpoint, glass—like any supercooled liquid—is in an unstable or metastable state. A metastable state is one of the possible states of a system, stable under the given conditions, which does not correspond to the maximum free energy (for example, a supercooled liquid, a supersaturated solution, etc.). A given phase can exist indefinitely in a metastable phase until a nucleus of another phase appears in it. It then passes spontaneously into a stable phase—a process accompanied by decrease of the free energy of the system. However, a certain amount of work is required for formation of the nucleus, and Gibbs takes this as the measure of stability of a metastable system.

Tammann [3], who studied crystallization of organic glassy substances in detail, distinguished two processes, which in general occur at different rates and in different temperature ranges: formation of nuclei (centers) of crystallization and crystal growth.

Despite a number of objections, concerned mainly with his techniques, Tammann's main principles are still regarded as valid [4].

The molecules in a liquid are in continuous random motion. The kinetic energy of the moving molecules decreases with fall of temperature, and accidentally formed aggregates become more stable. Finally, at a certain temperature, formation of crystal nuclei becomes possible.

The nucleation rate (number of nuclei appearing in unit time in unit volume of liquid) depends on the temperature of supercooling. At the melting point the crystals of a substance are in equilibrium with the melt, but nuclei are not formed. The number of nuclei increases with increase of supercooling, reaches a maximum, and then falls to zero. If a melt is cooled rapidly enough below the nucleation temperature, it solidifies in the form of an amorphous glass.

The lower the temperature of a supercooled liquid, the greater is its internal friction and the more difficult it is for the particles to move. This decreases the probability of nucleation, but the nuclei already formed become more stable. At a certain temperature, the conditions become most favorable for formation of crystal nuclei. The curve for the nucleation rate as a function of the temperature of supercooling has a maximum at that temperature.

The curve representing the variation of the linear rate of crystal growth with temperature also has a maximum. The course of crystallization depends on the relative positions of the nucleation and crystal growth-rate curves. Examples of different relative positions of these curves are shown in Fig. 1.

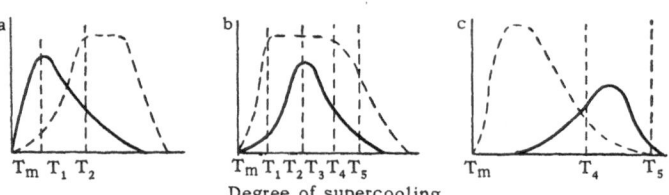

FIG. 1. Three types of relationships among the number of nuclei, the rate of crystal growth, and the degree of supercooling [3]: - - - - - rate of crystal growth; _____ number of nuclei.

In the first case (Fig. 1a), if the cooling is slow, numerous nuclei appear at the temperature T_1, corresponding to the maximum of the nucleation curve. On further cooling the crystals grow rapidly and a fine-grained polycrystalline substance is obtained. If the cooling is rapid, the region of maximum nucleation is quickly passed, and rapidly growing but not numerous nuclei appear at the temperature T_2, corresponding to the maximum rate of crystal growth. A coarse-grained polycrystalline substance is obtained as a result.

In the second case (Fig. 1b), the temperature of the maximum rate of crystal growth, T_1, is reached before the temperature corresponding to the maximum number of nuclei, T_2. If the cooling is slow, the number of nuclei at temperature T_1 is small; they grow rapidly, and a coarse-grained crystalline substance is formed. When the cooling is rapid, numerous rapidly growing nuclei appear at temperature T_2, and a fine-grained solid is obtained. If the cooling is even more rapid (supercooling T_2), the number of nuclei is small and growth is slow. If the glass is held at this temperature for a sufficiently long time, a coarse-grained material can be obtained; if the cooling is continued, individual grains in an amorphous glassy substance are formed.

In the third case (Fig. 1c), with deep supercooling, for example to T_4, the grains are not numerous and they grow slowly. If the substance is then cooled rapidly, individual grains will be retained in an amorphous solid mass. This happens during cooling of glass, when it contains rare individual crystals in the form of spherulites, stars, etc.

However, any kind of constant supercooling is rare in practice. Different rates of melt cooling are more common. The degree of supercooling is then altered; this complicates the process.

These concepts have been developed further in a number of recent publications on the crystallization of glass. The formation of crystallization centers in glasses, often described as nucleation, is examined by Stookey [5], Kleber [6], Hillig [7], and others.

In a paper which appears to be a generalization of previously published researches, Stookey [5] presents his theory of nucleation. Glass is considered as a supercooled liquid without any long-range order. (Instances of separation of a glass into two liquid phases are specifically distinguished as special cases.) Increase of the free energy of the system is required for formation of a crystallization nucleus. Excess free energy arises at the boundary between the liquid phase and the nucleus forming from it. For very small nuclei, with large surface area relative to their volume, the surface energy considerably exceeds the free energy released in the transition of a given amount of the substance from the liquid into the stable crystalline state. The excess and total change of free energy are positive. With increasing size of the nuclei, the surface energy becomes only a small part of the total change of energy, so that the total energy change becomes negative.

The change of free energy ΔF in formation of a spherical nucleus is represented by the equation

$$\Delta F = \frac{4}{3}\pi r^3 \Delta f_v + 4\pi r^2 \Delta f_s \tag{1}$$

where Δf_s is the change of free energy per unit surface area, Δf_v is the change of free energy per unit volume in the transition from one phase into the other, and r is the radius of the nucleus.

For certain values of Δf_s and Δf_v, with a certain critical radius r_{cr}, the change of free energy ΔF is maximal.

Increase of free energy is required for formation of particles of radius less than r_{cr}; particles of radius greater than r_{cr} grow with decrease of free energy and form stable nuclei.

The nucleation rate J is proportional to the probability of formation of stable nuclei

$$J = A \exp\left(\frac{-\Delta F^*}{kT}\right) \tag{2}$$

where ΔF^* corresponds to the maximum free energy when

$$r_{cr} = \frac{2\Delta f_s}{\Delta f_v} \tag{3}$$

$$\Delta F^* = \frac{\frac{16}{3}\pi(\Delta f_s)^3}{(\Delta f_v)^2} \tag{4}$$

The nucleation rate depends not only on the number of regions with energy states ΔF^* but also on the possibility of surmounting the energy barrier. The final approximate expression for the nucleation rate is

$$J \approx \frac{N^* kT}{h} \exp\left(-\frac{\Delta F^* + Q}{kT}\right) \tag{5}$$

where N^* is the number of atoms in unit volume, h is Planck's constant, k is Boltzmann's constant, T is the absolute temperature, and Q is the activation energy for diffusion.

The exponential term is predominant in this equation. At a low degree of supercooling, ΔF^* is large because Δf_v is small and the nucleation rate is not high. On further supercooling, Δf_v increases until ΔF^* becomes comparable to Q. This is the region of the highest rate of nucleation. On further decrease of temperature, ΔF^* becomes very small in comparison with Q and the nucleation rate falls. The size of the smallest stable nucleus decreases with fall of temperature.

Introduction of foreign particles into the system, the appearance of new phase boundaries, favors crystallization. In such cases the so-called heterogeneous nucleation occurs. Nuclei are formed on the foreign inclusions, at the interfaces, since the energy barrier (increase of surface energy) is less in such cases than in homogeneous nucleation.

This is the basis of coarse surface crystallization of glass, and also of the action of nucleation catalysts, such as salts of certain metals, in glasses. As soon as the first crystalline particles of the catalyst—metallic silver, gold, or platinum—have formed in the glass, they begin to act as heterogeneous nucleation centers.

The activation energy is usually lower for heterogeneous than for homogeneous nucleation; it is considered that in most cases crystallization of glasses is the consequence of heterogeneous nucleation.

If the degree of supercooling is considerable, the nucleation rate in glasses is low because diffusion to the nucleation centers is retarded by the high viscosity.

The relatively simple relationships derived theoretically for one-component systems become more complex in the case of multicomponent systems. When a system is supercooled with respect to several phases, the first to crystallize is the one with the lowest energy for activation of nucleation. Then, as the composition alters as the result of crystallization, consecutive deposition of several crystalline phases becomes possible. Because of the high viscosity of the melt, which hinders diffusion, deposition of equilibrium phases may not occur during the crystallization time, which is restricted in practice. Factors which favor decrease of the activation energy for nucleation accelerate and facilitate crystallization.

In Kleber's paper [6] nucleation is regarded as a kinetic process. The frequency of formation of nuclei depends on two factors: the work of nucleation proper and the activation energy for transfer of the substance to the nucleation site. The work of nucleation A_0 is the difference of two components—increase of energy on formation of a new surface, and decrease of energy in transition of the substance into the crystalline state. It can be estimated by Kaischew's method [8] from ψ, the work of separation of two neighboring elements in a two-dimensional model of the crystal lattice (Fig. 2a):

$$A_0 = \frac{\psi^2}{kTS}$$

where S is the degree of supercooling, T is the absolute temperature, k is Boltzmann's constant, and ψ is the work of separation.

The work of nucleation increases with decrease of the supercooling S. At a low nucleation rate (at low degrees of supercooling), the part played by the activation energy for diffusion increases. At zero supercooling the work of nucleation becomes infinitely large and nuclei are not formed.

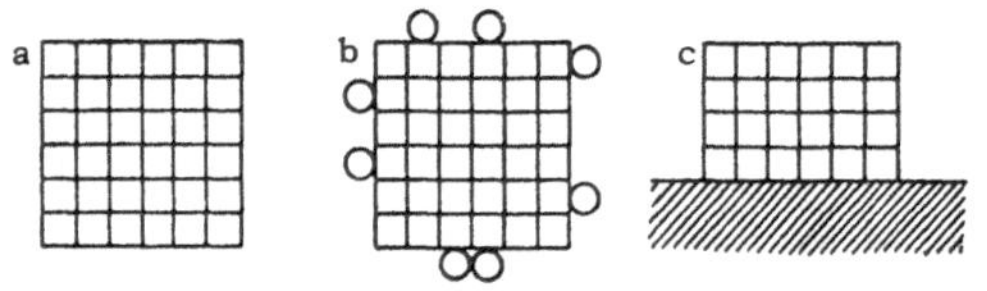

The size of a critical nucleus, like the work of its formation, depends on the supercooling.

Denoting by n_0 the number of structural elements on one side of a square nucleus (with a two-dimensional model), we have

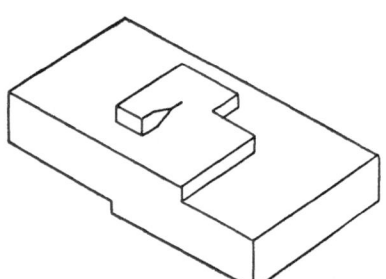

FIG. 2. Model of a crystal nucleus with a square two-dimensional lattice [8].

$$n_0 = \frac{\psi}{kTS}$$

It is shown that the work of nucleation is reduced by a number of factors. A similar two-dimensional model can be used for calculating the decrease of the work on absorption of foreign substances (Fig. 2b) and for formation of a nucleus on a support of another substance (Fig. 2c). In this last case the greater the work of separation from the support ψ, tne less is the work of nucleus formation on the foreign substance A_1:

$$A_1 = A_0 \left(1 - \frac{\psi_1}{\psi}\right)$$

This applies in particular to cases of structural similarity between the support and nucleus materials.

FIG. 3. Scheme of spiral crystal growth.

Kleber then examines subsequent growth and draws particular attention to the fact that deviations from regular crystal structure occur almost invariably. Crystals grow in a spiral fashion, and not by growth of planes (Fig. 3). In this case the energy relationships are different and growth is more rapid. Kleber also examines the theory of crystal growth put forward by Peibst [9]. In accordance with this theory, between the crystal with a regular lattice and the completely "random" melt there is an intermediate layer with gradually increasing structural order.

The general conclusion drawn from a series of experiments, in particular on the crystallization of phosphate glasses, is that crystallization of glasses does not differ in principle from crystallization of any melt. The process is determined by the relation between two basic quantities: activation energy for diffusion and work of nucleation.

Thus, at a high enough cooling rate, glasses can be obtained free from crystallization nuclei. However, such glasses can be crystallized if they are subsequently held in the temperature region favorable for nucleation. The time required for nuclei to form depends on the thermal history and the state of the system. Hillig [7] discusses this question in the light of Turnbull's theoretical considerations [10].

Consider a molten homogeneous glass which is cooled rapidly from above to below the liquidus temperature. The minimum time required for formation of a nucleus of the critical size can be estimated on the assumption that a given atom behaves as a sorbent, capturing other atoms which collide with it during random movement in the glass. The average time $\bar{t}$ for formation of an aggregate of critical radius r_{cr} is approximately represented by the equation

$$\bar{t} = \frac{\pi V_L^2 r_{cr}^2}{4 D V_M^2 x^2}$$

where V_M is the molar volume of the solid phase, V_L is the molar volume of the liquid phase, x is the molar fraction of the dissolved substance in the melt, and D is the coefficient of diffusion.

Substitution into this equation of typical values r_{cr} = 5 A, D = 10^{-17} cm^2/sec at a viscosity of 10^{10} poises, x ≈ 0.3, and $V_L = V_M$, gives $\bar{t} \approx$ 1 hour.

It is likely that this time is less than the time required to reach equilibrium conditions in practice because the instability inherent in nuclei close to the critical size was not taken into account.

The probability of an atom, in its random motion, entering the space between S and S + dS in time t is given by the equation

$$P dS = A \exp\left(\frac{-\tau}{t}\right) dS$$

where t is the holding time of the melt at constant temperature and τ is a characteristic transient time.

A similar expression is obtained for the nucleation rate

$$y(t) = y_0 \exp\left(\frac{-\tau}{t}\right)$$

where $\tau \geq \bar{t}$.

Since instantaneous cooling is impossible in practice, the transient time τ may have different values. The lower the cooling rate, the less is τ at a given degree of supercooling. If r_{cr} is of the atomic order of magnitude, $\bar{t}$ must be at least as large as the average time for a molecular transition.

Several quantities must be known for theoretical calculations of nucleation rates. There is as yet no independent method for measuring one of the most important parameters—the surface energy at a melt-solid interface. Rigorous methods for calculating the transient time are also lacking. In systems for which interfacial tension has been determined, nucleation obeys the theoretical laws. It is therefore possible to solve the converse problems—to determine σ by observation of homogeneous nucleation. Turnbull [10] found that σ_g, the molar free energy at the interface between relatively simple solids and their pure melts, is roughly proportional to the latent molar heat of fusion ΔH_f. The proportionality factor B = $\sigma_g/\Delta H_f$ is ~ 0.5 for simple monatomic metals and ~ 0.3 for more complex substances. This value can be used for finding the approximate value of the surface energy at the interface. For example, for a pure substance of melting point T_0 = 1500°K, molar volume V_M = 30 cm^3, and entropy change ΔS_f = 8 e.u., the surface energy at the interface is ~ 60 ergs/cm^2.

On the basis of these considerations Hillig plotted curves representing the dependence of the rate of homogeneous nucleation on the holding time of the glass at various temperatures. The shape of the curves depends on the liquidus temperature, surface tension, and viscosity of the glass. There is both a lower and an upper temperature limit for detectable nucleation at different holding times. Figure 4 shows variations of the nucleation rate in a glass with a liquidus temperature of 1500°K and $\sigma = 50$ ergs/cm^2 for different holding times. For example, if the

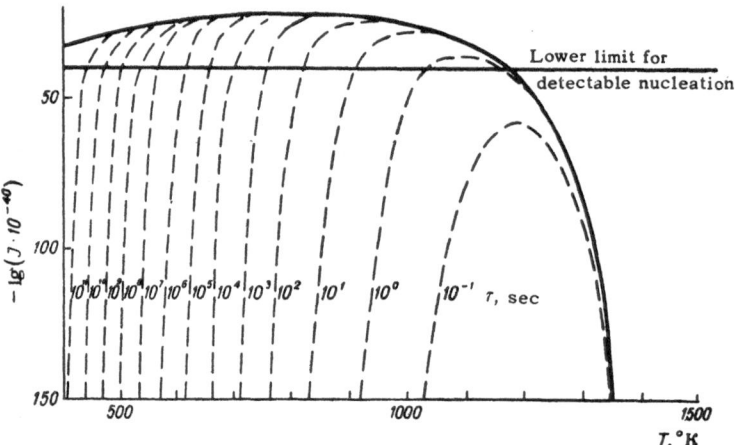

FIG. 4. Dependence of the nucleation rate on holding time [7].

glass is held for 1 sec, nucleation may be observed in the 1025-1150°K range. If it is desired to quench this glass without formation of nuclei, this temperature range must be passed in less than 1 sec. If the glass contains no crystallization centers, the holding time for detectable nucleation, e.g., at 760°K, must not be less than 10^3 sec.

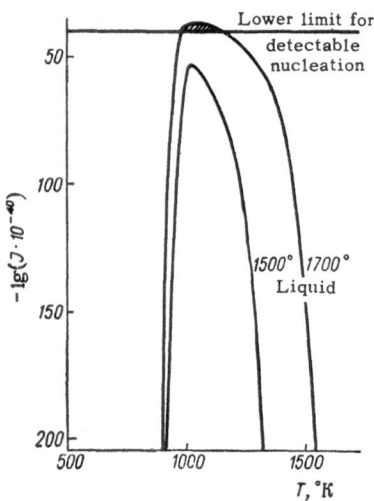

FIG. 5. Dependence of the nucleation rate on the liquidus temperature [7] (holding time $\tau = 10^3$ sec).

Figure 5 shows the influence of the liquidus temperature for glass with $\sigma = 63$ ergs/cm^2. Lowering of the liquidus temperature by addition of a component having no significant influence on viscosity or surface tension may intensify the stability of the system to homogeneous nucleation.

Effects of viscosity and surface tension on the kinetics of nucleation are shown in Fig. 6. The liquidus temperature is 1500°K in all cases.

Hillig discusses the question of heterogeneous nucleation in detail. In principle, any boundary between two phases one of which is the melt can catalyze nucleation. The better the wetting of the catalyst by the crystallizing substance in presence of the melt, the stronger is the action of the catalyst. The influence of a foreign substance (catalyst) on nucleation kinetics depends on the contact angle θ at the point where the support, melt, and the crystallizing substance meet. The work of formation of crystal nuclei in heterogeneous nucleation W^*_S is connected with the work of homogeneous nucleation W^*_0 by the expression $W^*_S = f(\theta) W^*_0$. The factor $f(\theta) = (2 + \cos \theta) \times (1 - \cos \theta)^2/4$. The angle θ is found from the surface tension values:

$$\sigma_{S,L} = \sigma_{1,S} + \sigma_{1,L} \cos \theta$$

where $\sigma_{S,L}$ is the surface tension at the support-melt interface, $\sigma_{1,S}$ is the surface tension at the crystal-support interface, and $\sigma_{1,L}$ is the surface tension at the crystal-melt interface.

Hillig stresses that not every phase in the melt assists heterogeneous nucleation. In a study of the crystallization of certain glasses in the systems $BaO - SiO_2 - TiO_2$ and $BaO - Al_2O_3 - SiO_2 - TiO_2$, the glasses were heated under a variety of conditions. On the basis of his experiments Hillig concludes that in these glasses homogeneous nucleation occurs in a definite range of compositions. In another region separation into two liquid phases is apparently possible. However, Hillig considers that separation into two liquid phases does not necessarily lead to nucleation—homogeneous or heterogeneous.

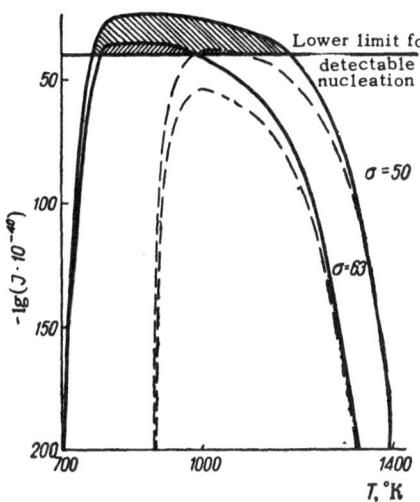

FIG. 6. Dependence of the nucleation rate on surface free energy and the coefficient of diffusion [7] (holding time $\tau = 10^3$ sec).

Many authors [11-14] consider that separation into two glassy microphases precedes and favors uniform crystallization of glass throughout its volume with formation of small crystals. This question is discussed in detail by Hinz and Kunth [11].

Separation into two immiscible liquid phases is observed in many systems. If it occurs above the liquidus temperature, it is known as detectable separation. In some systems separation occurs in latent form, below the liquidus temperature, i.e., in the supercooled state. In the former case the phase diagram of the system has a region of two immiscible phases above the liquidus line, while in the second the liquidus curve is characteristically S-shaped.

Many authors have shown that it is possible to calculate the limiting concentrations of components forming separating melts from data on the geometrical dimensions of the atoms and ions, their coordination numbers, bond strengths, polarizability, etc.

Warren and Pincus [15] were the first to apply the principles of crystal chemistry to the problem of phase separation. On the basis of a study of binary borate and silicate systems, they showed that tendency to phase separation is associated with the tendency of modifier and network-former cations to surround themselves with oxygen anions in accordance with their coordination numbers. The stronger the bond between the modifier cation and oxygen, the greater is the tendency for an independent phase enriched with the modifier to separate from the melt. The size and valence of the ions play a part here.

Dietzel [16] considers that the quantity Z/R^2, where Z is the valence of the cation and R is the distance between the cation and oxygen (the sum of the ionic radii), is a measure of the tendency to separation. The greater this quantity, the greater is the tendency to separation.

These points are discussed in detail by Levin and Block [17]. They showed, for various binary and ternary borate, silicate, and borosilicate systems, that it is possible to calculate the limiting concentrations of modifiers favoring separation when introduced into melt. In silicate and borate systems one of the separating phases nearly always contains ~100% of the network-forming oxide. The other phase is enriched with the modifying oxide. The limiting composition—the phase composition with the highest modifier content—can be calculated with sufficient accuracy from data on density, coordination numbers, and types of coordination.

In a given system the network-former and modifier tend to form their own coordination polyhedrons, surrounding themselves with oxygen anions in accordance with the ratio of the ionic radii. If the oxygen polyhedrons around the modifying cations are similar in type and size to the polyhedrons forming around the network-forming cations, phase separation does not occur. If the polyhedrons are of different types, there are different possibilities, dependent on the electrostatic bond strength Z/CN (where Z is the valence and CN is the coordination number): open or latent separation (the latter occurs when the system is held below the liquidus temperature); the liquidus curve becomes S-shaped.

In silicate systems silicon always has a coordination number of four, and forms SiO_4 tetrahedrons with oxygen anions. On introduction of modifying oxides, oxygen is linked with Si^{4+} cations, with rupture of Si-O bonds. Two "nonbridge" oxygen ions are formed, while two SiO_4 groups acquire additional negative charges which are balanced by the positive modifying ions. The modifier cations may be attached to the same nonbridge oxygen ion at an angle of 180°C, when the distance between them becomes $2(1.40 A + R_{cat})$, since the radius of the oxygen ion is 1.40 A. This coordination type is termed coordination type A by Levin and Block. In type B coordination the cations are attached to opposite sides of an SiO_4 tetrahedron at opposite pairs

9

of oxygen ions, and this balances the total negative charge of the tetrahedrons. The two coordination types are shown schematically in Fig. 7. In type B coordination the distance between the cations is

$$S = 1.87 + 2\sqrt{(1.40 + R_{cat})^2 - 1.75}\,\text{Å}$$

for silicates, and

$$S = 1.77 + 2\sqrt{(1.40 + R_{cat})^2 - 1.57}\,\text{Å}$$

for borates; i.e., greater than in type A coordination. Therefore, a change of coordination type from B to A leads to an increase of the amount of the modifying oxide in the liquid phase enriched with the modifier. Other changes of the coordination type have similar effects. For example, in a number of binary borate and silicate systems the bivalent cations have type B coordination. In ternary borosilicate systems the coordination changes to type A, which increases immiscibility.

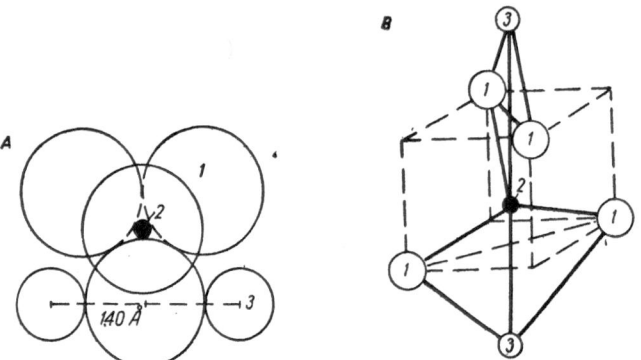

FIG. 7. Coordination types of modifier cations [17]: 1) oxygen; 2) network-forming ion; 3) modifier.

Levin and Block give a table from which conclusions may be drawn concerning the influence of the type of coordination and of electrostatic bonding strengths between cations and oxygen on immiscibility (Table 1).

When the electrostatic bond strength between the modifier and oxygen is greater than $\frac{1}{4}$, separation into two liquids occurs. No visible separation occurs when the bond strength is less than $\frac{1}{4}$. However, the one-phase system formed is unstable, and this influences the shape of the liquidus curve.

Modifier ions with bond strengths of $\frac{1}{4}$ or less have type B coordination, and those with bond strengths of $\frac{1}{3}$ or more have type A. The change of the coordination type with change of bond strength is explained by different charge distributions on the tetrahedrons, in accordance with Pauling's principles. The maximum possible excess charge on each oxygen ion, with uniform distribution between all the ions in the tetrahedron, is $\frac{1}{4}$. Therefore, type B coordination is impossible for cations with bond strength exceeding $\frac{1}{4}$, whereas univalent ions, for example, always give type B coordination.

Glasses with different types of coordination differ in structure. The authors specially emphasize that even glasses with the same type of coordination differ in structure because of differences in coordination numbers, valences, and amounts of oxygen ions introduced. For example, potassium and lithium have the same valence but different coordination numbers; potassium and barium have the same coordination number but different valences. Addition of modifying oxides to borate glasses first leads to a structural change, with alteration of the coordination number of boron from 3 to 4, whereas in silicate glasses rupture of Si-O bonds occurs at once.

With cations having weak electrostatic bonds the liquidus line is straight, and with increase of bond strength the liquidus curve becomes S-shaped (for example, in the case of sys-

TABLE 1
Electrostatic Bond Energies Between Cations and Oxygen, Z/CN, Coordination Types, and Form of Liquidus Curves for Certain Binary Silicate Systems [17]

Cation	Cation radius (after Pauling) R_{cat}, Å	Ratio $R_{cat}/R_{O_2} = R_{cat}/1.40$	Coordination number, CN	Electrostatic bond energy, Z/CN	Form of liquidus curve	SiO$_2$ coordination type
Cs$^+$	1.69	>1	12	$1/12$	Linear	Not determined
Rb$^+$	1.48	>1	12	$1/12$	»	Not determined
K$^+$	1.33	0.95	8	$1/8$	S-shaped	B
Na$^+$	0.95	0.68	6	$1/6$	»	B
Li$^+$	0.60	0.43	6	$1/6$	»	B
Ba^{2+}	1.35	0.96	8	$1/4$	»	B
Pb^{2+}	1.21	0.86	8	$1/4$	—	B
Sr^{2+}	1.13	0.81	8	$1/4$	2 liquids	B
Ca^{2+}	0.99	0.71	6; 8	$1/3$; $1/4$	2 »	B
Cd^{2+}	0.97	0.69	6	$1/3$	2 »	A
Fe^{2+}	0.75	0.54	6	$1/3$	2 »	A
Zn^{2+}	0.74	0.53	6	$1/3$	2 »	A
Co^{2+}	0.72	0.51	6	$1/3$	2 »	A
Mg^{2+}	0.65	0.46	6	$1/3$	2 »	A
Be^{2+}	0.31	0.22	4; 3	$1/2$; $2/3$	1 liquid	—
Cr^{3+}	0.64	0.46	6	$1/2$	2 liquids	—
Ti^{4+}	0.68	0.49	6	$2/3$	2 »	—
Network-forming ions:						
B^{3+}	0.20	0.14	3; 4	1, $3/4$	—	—
Si^{4+}	0.41	0.29	4	1	—	—

tems with K and Li). When the bond strength is $1/4$ two liquids can be formed. Finally, all cations with bond strengths of $1/3$ give rise to two liquids in systems with SiO$_2$ and B$_2$O$_3$.

Filipovich [13] also put forward a "statistical-thermodynamic" theory of formation of new phases in glass melts. Two groups of relaxation processes, leading to establishment of stable or metastable equilibrium, are possible, dependent on the composition of the glass, the melt temperature, and energy relationships.

The first group comprises processes leading to formation of a new glass structure, for example, separation into two new glassy phases or metastable separation at temperatures below the solidus. The second group consists of relaxation processes of crystallization. Processes of both groups may occur simultaneously, being superposed.

Filipovich's article gives formulas and graphs for the dependence of the free energy of a two-component system on composition for various types of phase separation and crystallization. In some glasses metastable microseparation may appear and disappear reversibly without crystallization occurring. In other instances the compositions of the newly formed glassy phases resulting from separation approach to stoichiometric; this facilitates the initial stages of crystallization and determines to some extent the dimensions of the future crystals.

The following conditions are essential for formation of uniform fine crystals in the glass: 1) nuclei must form throughout the volume of the glass, 2) the number of nuclei formed when the glass is held below the crystallization temperature must be fairly large—of the order of 10^{12} per cm^3 or over, and 3) the surface energy at the interface between the crystals and the glassy interlayer must be low to ensure good wetting.

Under these conditions "the glass itself automatically selects, by virtue of the principle of maximum thermodynamic potential and the laws of statistics, the crystallization route and the sequence of phase deposition which ensure the minimum mechanical stresses in the glass and the minimum number of fissures, i.e., the maximum strength in the product of crystallization" [13].

Filipovich points out that crystallization of glasses exhibiting phase separation cannot be described as "catalytic," as is done by many authors. The matter is not one of catalysis but of the influence of phase separation on the subsequent course of crystallization. It is possible that the main function of metastable separation is to create small glassy regions, the dimensions of which determine the order of magnitude of the crystals formed, rather than to increase the interfacial area. On the basis of various considerations, the author concludes that "if precrystallizational separation occurs, it must be of eutectic character with subsequent crystallization of stoichiometric glassy nuclei. Moreover, it may be concluded that if metastable separation of the eutectic type, if detected in a given glass, must always be of the precrystallization type; once it has occurred, it does not disappear and culminates in crystallization of the glass at all temperatures below the solidus line. This question requires separate experimental investigation."

Hinz [11] emphasizes that on energy considerations separation into two liquid or glassy phases should occur more easily than formation of crystal nuclei in a glassy phase. This is because the interfacial energy between two liquids is very small if the difference between their surface tensions is not large. The interfacial energy between a glassy phase and the crystals separating from it is usually greater.

After separation has occurred, regions containing particles uniform in composition are formed. Therefore the work of diffusion for their coalescence is not large either. The crystallization tendencies of the separated phases are different; one may crystallize more rapidly than the other. Special heat-treatment procedures are necessary if separation is to lead to subsequent uniform crystallization; the microheterogeneity regions must be of a definite minimum size.

Roy [18, 19] examines the relation between the phase diagram of a system and the ability of the system to undergo uniform volume crystallization. The formation of a multitude of crystallization nuclei can be predicted from the form of the phase diagram. For example, in silicate systems which separate in the liquid phase, separation in the metastable state is also possible. Phase diagrams of such binary systems are shown schematically in Fig. 8. The hypothesis is advanced that the structure of the liquid adjacent in composition to the field of equilibrium between two immiscible macrophases should be of a specific character. It probably consists of two types of glass distributed in each other. The size of the heterogeneities is of the order of 10-100 A. One of the glasses is richer in silica and the other in the modifier ion. In a state of thermodynamic equilibrium a continuous exchange occurs between these heterogeneous regions, which have no distinct interfaces. The further the composition from the separation region, the more homogeneous is the glass and the smaller are the heterogeneities. In a glass with a fine heterogeneous structure the course of crystallization may differ in accordance

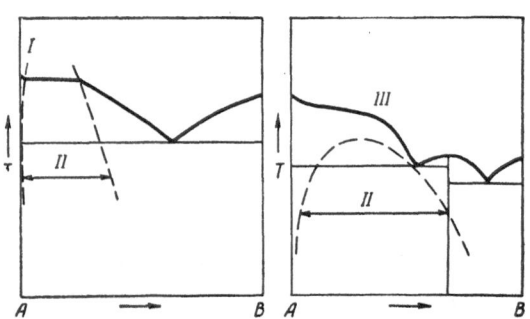

FIG. 8. Phase diagrams of binary systems capable of metastable separation during heat treatment below the solidus line: I) Two stable liquids; II) two metastable liquids; III) liquids with special structure.

with the composition and conditions. A liquid which is supercooled as a transparent glass may be appreciably separated into two phases during subsequent heat treatment. The separation of a second liquid phase differs from crystal formation. The composition of the liquid phase may be entirely different from that of the possible crystals. The rate of formation of the second liquid phase may be considerably higher than the crystallization rate, because, unlike crystallization, it involves displacement of whole blocks of the same composition, without rupture of cation-oxygen bonds. The activation energy for separation of a liquid phase is less than the activation energy for crystallization. Therefore separation into liquid phases may occur at lower temperatures. The compositions must be close to the separation region, or must give liquidus lines of flattened form, which indicates the same tendency to separation, but to a lesser extent.

Roy [19] stresses the indefinite nature of the "glass" concept. Glasses of the same composition may be entirely different in structure and free energy, in accordance with the initial cooling conditions, subsequent heat treatment, etc. A glass always differs to some extent from the state of an ideal supercooled liquid. Figure 9 represents variations of the free energy F of a system with increase of supercooling. Different cases are possible during isothermal crystallization of glass. The original glass may be in different states. For example, it may have a fairly pronounced random liquid structure (glass 1), a more orderly structure (glass 2), or, in the extreme case, separation into two phases with short-range order may occur (glass 3). Each glass has its own "partial" curve for the dependence of free energy on temperature within a certain temperature range. An ideal supercooled liquid is the extreme case of complete reversibility of the change of free energy with temperature. In other cases the changes depend on the time and temperature and take different courses, dependent on the initial state.

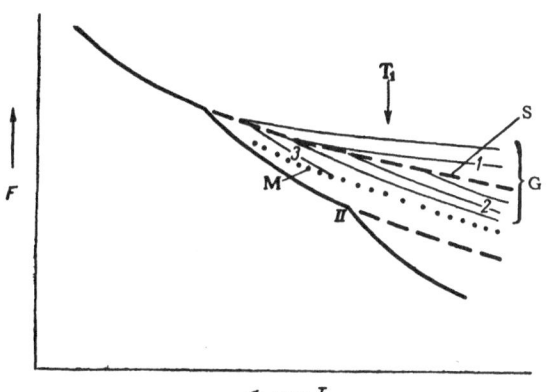

FIG. 9. Schematic representation of the change of free energy F of a system with temperature T [19]: region above the point I, for the liquid; region I–II, for crystals A; region below the point II, for crystals B; $G_{1,2,3}$, for glasses of the same composition but different degrees of structural order; S, for an ideally supercooled liquid; M, for the existence of two metastable liquids in the system.

Figure 10 shows the course of crystallization of a glass with composition A at constant temperature T_1. The glass is quenched at a high temperature, heated up again to temperature T_1, and held at that temperature. Complete equilibrium is reached at the point E, where the glass is fully crystallized. However, various intermediate stages are possible between the points A and E. Any glass must pass through the region AB, where the free energy is less than free energy of an ideal supercooled liquid. If the system consists of two (or more) components, separation into two liquid phases is possible at a certain intermediate stage C. The structural changes in the region A-B are small (they may be regarded as second-order changes) and occur continuously. If they do not influence the volume, refractive index, or density, they may even escape notice. Second-order changes may also occur in the crystallization region (represented by the dash line above the curves for the crystalline phases). Formation of crystallization nuclei is not the most stable state. The surface free energy decreases with growth of the crystals and gradually falls to a value corresponding to the F versus T curve for the crystalline phase. Thus, the stages enumerated below, which may in practice be superposed on each other and complicate the general picture, can be distinguished in the crystallization process: 1) continuous second-order changes in the glassy phase; 2) possible formation of centers of a second short-range order phase, which restricts structural changes in the first phase; 3) formation of nuclei of a metastable crystalline phase; and 4) formation of nuclei of a stable crystalline phase, and their growth as the result of reactions or transformations of metastable crystals.

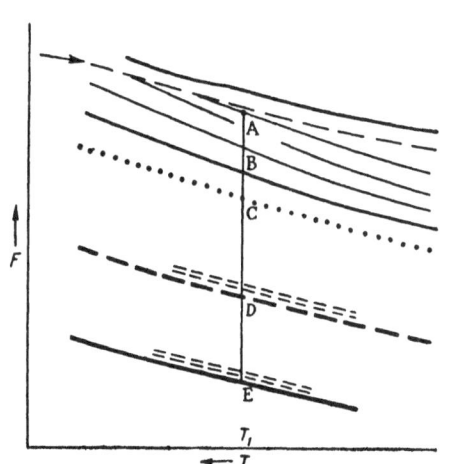

FIG. 10. Part of the F versus T diagram illustrating isothermal crystallization of glass heated to state A [19]: - - - - - ideally supercooled liquid; . . . . two metastable liquids; - - - - metastable crystals; ——— stable crystals.

Formation of two immiscible liquid phases, which occurs in a number of silicate systems, may give rise to heterogeneities of the order of 20-100 A without considerable expenditure of energy. The tendency of any liquid to separate into two liquid phases may be intensified by suitable treatment. Potential nuclei of less than critical size exist in large numbers in the liquid. These are regions differing slightly in composition, in dynamic equilibrium with each other, without interfaces. Quenching or repeated heating may cause the centers to grow, since the activation energy for the process is low and the nuclei are so numerous that diffusion needs to occur over only very short distances. It is impossible to characterize the process quantitatively at this stage, because the surface energy, stress energy, and composition fluctuations are indeterminate. It may be supposed, however, that extremely fine homogeneous nucleation occurs, starting with separation of a second phase with short-range order.

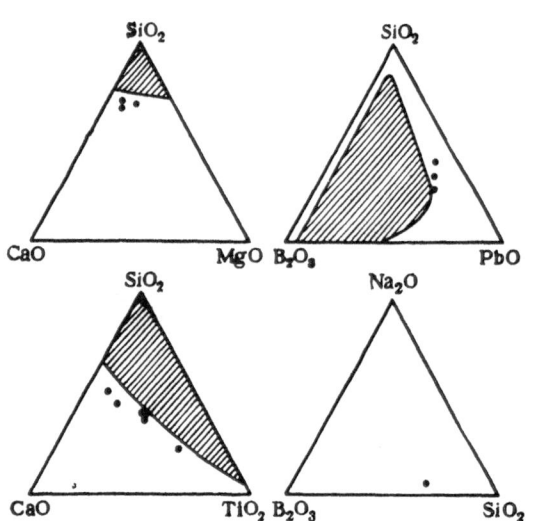

FIG. 11. Systems for which the possibility of homogeneous nucleation was predicted from the form of the liquidus curves [19].

Homogeneous nucleation occurs in systems characterized by separation (open or latent) into two liquid phases. Phase separation precedes crystallization. Since both glassy phases are metastable, crystallization is accompanied by decrease of free energy. Roy refers to several systems with phase diagrams indicating phase separation, either detectable or latent, revealed by heat treatment. Such behavior is most likely in glasses close to the immiscibility regions (detectable) and glasses with latent regions of separation (Fig. 11). Composition regions in which separation occurs are hatched in the diagrams. The points indicate glasses in which homogeneous nucleation occurred, as the result of microheterogeneity, after repeated heating and holding at temperatures several hundred degrees below the liquidus temperature. Roy's views are confirmed by a number of investigations in which it was possible to demonstrate clearly that separation into glassy phases is a stage preceding crystallization. These investigations include the work of Vogel, Maurer, Ohlberg [20], and others.

Vogel and Gerth [12] examine glass crystallization in the light of the data of various investigators [20-28] on the presence of microheterogeneous regions in glass. They used fluoberyllate glasses, which can serve as models of silicate glasses. In these glasses all processes occur at lower temperatures, are more distinct, and are easier to study. It is difficult to detect heterogeneity in ordinary silicate and borate glasses, because the heterogeneity regions are small and are beyond the resolving power of the electron microscope. On the basis of their experimental data, they propose the following picture of glass structure.

"On cooling a liquid glass melt, aggregational processes of atoms or atom groups set in, caused by the same forces that under normal conditions are responsible for a regular crystal structure. These aggregational orientations remain fixed at an intermediate stage between ideal disorder of the material components in the melt and the crystalline state during the process of solidification. That is, clusters of homogeneous ions will aggregate in certain areas, thus forming the preceding and prerequisite state for crystallization.

"Inhibition of the orienting process by rapidly cooling the melt leads to glassy amorphous congelation of these areas, which under certain conditions will separate from one another in the form of droplets of aggregating clusters and glass, which develop crystallinelike phases. In such cases the structure of one vitreous phase may be better developed than that of the other."

The authors describe this type of glass structure as "cellular."

One of the phases is richer in modifying ions; the other, in network-formers. The phase with the greater surface tension forms droplets. Heat treatment plays an important part. The orientation process, interrupted by sharp chilling, may be renewed as the result of suitable heat treatment. Intensive separation may occur when a clear glass is held at a definite temperature. The drops grow and cause opalescence (for example, in borosilicate glasses). The temperature conditions may be so chosen that the drops do not exceed a certain optimum size and are distributed uniformly.

In a study of phase separation of model fluoberyllate glasses, Vogel concluded that microheterogeneity regions exist in all glasses. High viscosity and rapid cooling prevent complete separation of the glass during the initial cooling. The tendency to separation depends on the cation field strength. The higher the field strength of the modifier cation, the greater is the order in the phase in which it is in excess by comparison with the second phase, and the greater is the tendency of that phase to crystallization. An interface is formed if the difference between the surface tensions of the separating phases is large enough. Additives which influence surface tension, e.g., by equalizing the surface tensions of the two phases, decrease the tendency to separation.

When a glass separates into two glassy microphases, crystallization begins in regions with the more favorable conditions and of higher order. Usually this occurs within the drops.

Vogel gives a series of electron micrographs of silicate glasses containing TiO_2, and of fluoberyllate glasses, at various stages of crystallization (Fig. 12). In both cases the process begins with separation into liquid phases. The glassy phase richer in modifier cations usually

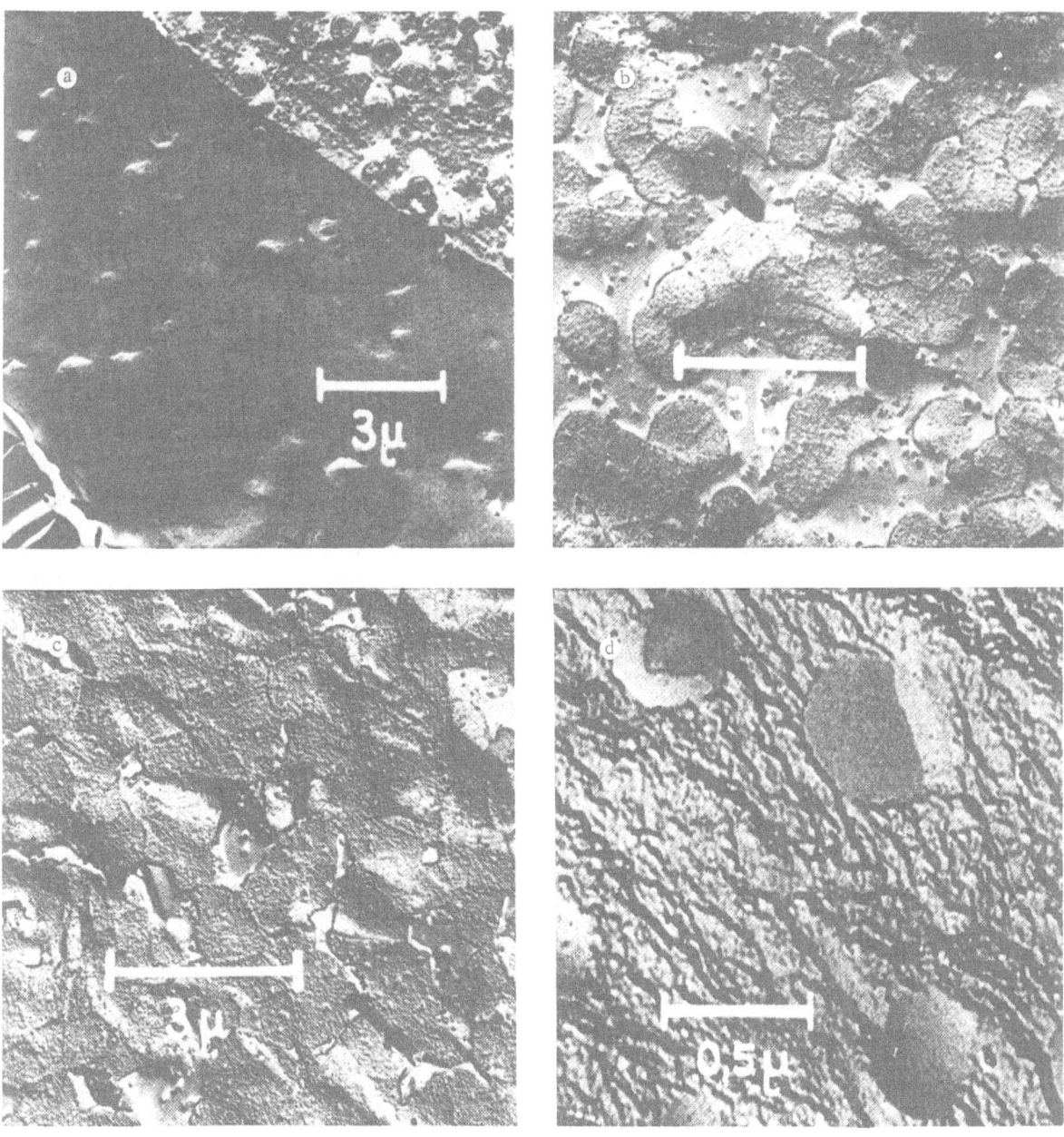

FIG. 12. Electron micrographs [12]: a) Annealed glass-ceramic with separation into droplets. The glass region in the top right corner is slightly etched with HF. b) Intensifying separation and crystallization as the result of heat treatment of the glass-ceramic. Many particles of the separated phase are in contact. This stage indicates the end of crystal growth. The specimen is slightly etched with HF. c) Further stage in the heat treatment of the glass-ceramic. Crystallization is almost total. The specimen is slightly etched with HF. d) Glassy droplike regions enriched with BeF_2 in a completely crystallized main phase.

crystallizes first. In this phase there are more discontinuities in the network and the ions are more mobile; it also usually contains the catalyzing ions. In silicate glasses the modifier-rich phase separates out in the form of droplets distributed in the phase rich in the network-former; accordingly, crystallization begins in the droplets. In contrast to silicate glasses, crystallization in fluoberyllate glasses begins in the phase surrounding the droplets enriched with the network-former.

Titanium dioxide is used in many glasses as a component favoring uniform volume crystallization. It is considered that TiO_2 helps separation of the glass on the one hand, and may form the primary crystalline phase serving as the heterogeneous nucleating agent on the other.

In discussing phase separation in the ternary system $Na_2O - B_2O_3 - SiO_2$, Vogel refers to the possibility of secondary conversions leading to formation of three-phase glasses. A second heat treatment of glasses in this system results in the formation of a glassy phase enriched with B_2O_3, with two other phases distributed in it in the form of droplets: one rich in SiO_2 and the other rich in borates. Addition of P_2O_5 to glasses in this system produces opacity. If CaO is also introduced into such a glass of definite composition, the drops increase considerably in size and secondary separation occurs within the drops themselves. Fine droplets containing increased amounts of P_2O_5 and CaO are formed within the drops of the phase rich in P_2O_5. Secondary separation also occurs during formation of glass-ceramics [29].

Vaisfel'd, Shelyubskii, Solomin, and Sorkin [30-33] investigated processes of glass crystallization by electron microscopy. Investigations of the structure of the original titanium-containing glasses showed that they are all microheterogeneous, with microheterogeneity regions of about 0.01μ. It was shown that crystallization in these glasses is due to metastable phase separation; growth of the separation regions during precrystallizational heat treatment was demonstrated.

Electron microscope studies of original glasses in the $Li_2O - SiO_2$ system, containing TiO_2 as the crystallization catalyst, showed the presence of numerous rounded regions of phase separation; these regions increase in size with increasing TiO_2 content. X-ray structural analysis showed that lithium disilicate and cristobalite separate out during subsequent crystallization of the glass. No phase associated with TiO_2 was detected.

Growth of the microcrystals at the initial stages in the formation of glass-ceramics is effected by diffusion of the substance corresponding to the composition of the growing crystalline particle from the surrounding glassy phase toward the nucleus. The growing microcrystals are surrounded by regions poorer in the material corresponding to the composition of the growing crystal. The size of these regions or "enclaves" of crystallization is associated with the size of the crystals growing in them: within a given glass the larger crystals are surrounded by larger crystallization regions. The shape of the "enclaves" correspond to the shape of the growing crystals when the latter are of considerable size.

Apart from electron microscope studies, the formation of microheterogeneities of the emulsion type in glasses during heat treatment has been confirmed by other methods.

Maurer [14] used the light-scattering method for studying the mechanism of nucleation in glasses of the system $MgO - Al_2O_3 - SiO_2$ with added TiO_2. Particular attention was devoted to dissymmetry and depolarization of the scattered light. The glasses were subjected to various heat treatments. The degree of crystallization of magnesium titanate was determined from the intensity of X-ray diffraction. The crystal size was estimated from the X-ray diffraction line broadening. The investigation showed that the glass contained heterogeneities even before heat treatment. However, depolarization data indicated that these heterogeneities are optically isotropic at first. Anisotropy gradually increases during the subsequent heat treatment; this may be attributed to crystallization within these regions, which are enriched with titanium dioxide.

Titanium dioxide tends to form aggregates in the amorphous regions, and these aggregates become more resistant to crystallization than the homogeneous glass. Crystallization begins in these regions.

As was stated earlier, the heat-treatment procedure as well as the composition of the glass is of great importance in crystallization of glass. Kitaigorodskii [34] investigated glasses in the system $MgO - Al_2O_3 - SiO_2$ with additions of TiO_2, SnO_2, ZrO_2, PbO, and F, and demonstrated the influence of heat treatment of the glass at a temperature below the crystallization temperature in the so-called "precrystallization period." It was shown by differential thermal analysis, X-ray phase analysis, and electron microscopy that preliminary heat treatment alters the character of the crystallization process, the phase composition, and the formation sequence of the crystalline phases, and makes it possible to obtain smaller crystals. Treatment of the glass at a temperature near the softening point has an especially strong influence. After treatment for 2 hours at this temperature, and then while being heated to 950°C, the glasses exhibited very fine crystallization, and the glass-ceramics formed remained transparent. The properties and structure of the crystallization products formed at higher temperatures were different, regardless of the temperature conditions in the precrystallization period. The temperature of the preliminary treatment can be lowered if the time is increased. Figure 13 shows the influence of temperature on the time required for heat treatment of the glass in the precrystallization period, ensuring formation of a transparent glass-ceramic during the subsequent treatment.

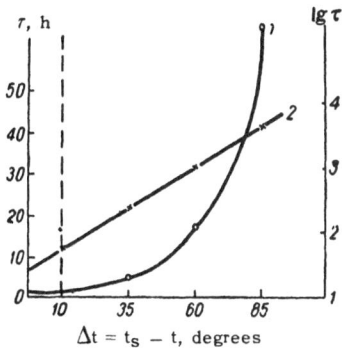

FIG. 13. Effect of temperature on the time τ of preliminary heat treatment required for production of transparent glass-ceramics [35]: 1) τ versus Δt; 2) lg τ versus Λt; t = temperature of heat treatment; t_s = softening temperature, °C.

It was shown by X-ray phase analysis and by electron microscopy that the glassy structure becomes more ordered during the precrystallization period, but the crystalline phase does not yet separate out. Additional treatment at higher temperatures is required for its separation.

Rabinovich and Sil'vestrovich [35-37] carried out a series of studies on crystallization of certain glasses under the influence of additives. The following problems were studied: 1) the formation mechanism and form of the particles of the second phase which are produced in glasses containing compounds of fluorine and phosphoric anhydride; 2) effects of fluorides, phosphoric anhydride, small amounts of metals, and other modifiers on crystallization of glasses (in particular, the role of crystallographic correspondence); and 3) course of crystallization of a number of glasses (mainly containing titanium), and methods of selecting the crystallization conditions.

Toropov [38] emphasizes that in most cases silicate melts are multicomponent systems, and after crystallization they are also multiphase systems, which complicates the course of crystallization considerably. Toropov examines various cases of crystallization of multicomponent systems. The simplest is eutectic crystallization of a three-component system. The following separate out successively: 1) the primary phase; 2) two phases in reciprocal eutectic relationships; 3) a triple eutectic. "One-step" crystallization is also possible; this is characterized by separation of crystals of the primary phase around the nuclei formed at the first stage, without formation of new nuclei of the primary phase at the second and third stages of the process. The author demonstrates, for crystallization of products made from blast-furnace slags, cast stone, and other silicate melts, that structures differing sharply from each other are formed, dependent on the course of crystallization.

Toropov [38] attaches great importance to the phase diagram of the system: "Whereas the formation of nuclei or crystallization centers is itself determined mainly by the structural conditions and to a certain extent by the correspondence between the motifs of various planar networks in the space lattices of the crystalline phase and nucleus, in the subsequent course of the crystallization process the decisive role belongs to the phase relationships of the system which are represented by its phase diagram and determine the character of the processes taking place in the system" [38].

Kleber [6], Stookey [5], and others refer to the influence of crystal-chemical similarity on crystallization of glasses.

Rabinovich [39] investigated the effect of added metals on crystallization of certain binary or ternary glasses, the primary phases of which are known from the phase diagrams. When there is crystal-chemical similarity between the nucleus and the crystallizing substance, i.e., when their linear lattice parameters (or multiples of them) are within 10% of each other, crystallization is accelerated considerably and occurs at lower temperatures than in absence of nuclei. Rabinovich states that the substances introduced into glasses in order to produce a uniform crystalline structure can be subdivided into two groups. Group A consists of substances having low solubility in glass, such as metals. The crystal size in the glasses may be of the order of 20-100 A. The cell constants of these metals are the same in glass as in the metals in the massive state. The cubic cell constant is 4.078 A for silver, 3.91 A for platinum, and 3.61 A for copper. Formation of lithium disilicate and cristobalite crystals is accelerated on silver crystals, of tridymite crystals on copper crystals, etc. Crystals of fluorides may play the same role as metal particles. Substances such as TiO_2, ZrO_2, etc., are assigned by Rabinovich to Group B. These substances have considerable solubility in glass, and, therefore, in distinction from substances of group A, they are added in amounts of 5-20%. The accelerating effect of group B substances on crystallization is hardly likely to be due to crystal similarity. Such oxides rarely separate out as the primary phase. Their salts should not crystallize more rapidly than silicates, so that it is improbable that they act as nuclei. The function of these substances appears to be that they cause immiscibility; "introduction of group B oxides makes it possible to bring almost any glass composition into the region of metastable separation. The subsequent crystallization is due to the very extensive interface between the metastable phases."

Weyl [40] examines immiscibility in silicate melts containing TiO_2 in the light of the theory of "shielding" of Si^{4+} and Ti^{4+} cations by oxygen anions. The smaller Si^{4+} ion (r = 0.41 A) is completely shielded by four oxygen ions; the larger Ti^{4+} ion (r = 0.68 A) requires sixfold coordination up to the melting point of rutile. However, if the oxygen ions are capable of strong polarization, a certain equilibrium is possible between TiO_6 and TiO_4 groups. The cations tend to decrease their coordination numbers with rise of temperature. Cooling of melts which do not contain O^{2-} ions of high enough polarizability to retain Ti^{4+} ions in fourfold coordination results in separation of a phase rich in TiO_2, in which Ti^{4+} ions are in sixfold coordination. In contrast to SiO_2, titanium dioxide has a defect structure with vacant anion sites. This leads to disproportion of the bonding forces and lowers the activation energy for nucleation. Therefore, TiO_2 separates out with formation of numerous nuclei.

In the publications cited in this survey the crystallization process is examined from various standpoints. Thermodynamic considerations are advanced and the parts, played by phase separation, crystal similarity, and the phase relationships in the phase equilibrium diagrams are discussed in detail. However, as yet no complete theory of crystallization has been advanced. The problem of controlled volume crystallization is highly complex and requires further investigations, which are of great practical and theoretical interest, since they lead to deep insight into the nature of the glassy state.

LITERATURE CITED

1. E. Ya. Mukhin and N. G. Gutkina. Crystallization of Glasses and Methods for Its Prevention, Oborongiz (1960).
2. A. I. Berezhnoi. Photosensitive Glasses and Crystalline Glass Materials of the "Pyroceram" Type, VINITI (1960).
3. G. Tammann. Z. Anorg. Chem. 181:408 (1929).
4. V. D. Kuznetsov. Crystals and Crystallization, Gostekhteoretizdat (1954).
5. S. D. Stookey and R. D. Maurer. Progress of Ceramic Science, Vol. 2, Oxford (1962), pp. 77-101.
6. W. Kleber. Silikat Tech., No. 1:5-10 (1962).
7. E. B. Hillig. Symposium on Nucleation and Crystallization in Glasses and Melts (1962), pp. 77-90.
8. R. Kaischew. Fortschr. Mineral. 38:7 (1960).

9. H. Peibst. Z. Phys. Chem. 216:304 (1961).

10. D. Turnbull. J. Appl. Phys. 21(10):1022-1027 (1950).

11. W. Hinz and P. O. Kunth. Silikat Tech. 11(12):506-511 (1960); W. Hinz and P. O. Kunth. Glastech. Ber., No. 9:431-437 (1961).

12. W. Vogel and K. Gerth. Symposium on Nucleation and Crystallization in Glasses and Melts (1962), pp. 11-22.

13. V. N. Filipovich. in: The Glassy State, Part 1, Izd. Akad. Nauk SSSR (1963), pp. 9-23. [English translation: The Structure of Glass, Vol. 3, Consultants Bureau, New York (1964), p. 9.]

14. R. D. Maurer. J. Appl. Phys. 33(6):2132-2139 (1962).

15. B. E. Warren and A. G. Pincus. J. Am. Ceram. Soc. 23(10):301-304 (1940).

16. A. Dietzel. Glastech. Ber. 22:41-50, 81-86, 212-224 (1948).

17. E. M. Levin and S. Block. J. Am. Ceram. Soc. 40(3):95-106; No. 4:113-118 (1957).

18. R. Roy. J. Am. Ceram. Soc. 43(12):670-671 (1960).

19. R. Roy. Symposium on Nucleation and Crystallization in Glasses and Melts (1962), pp. 39-46.

20. S. M. Ohlberg, H. R. Golob, and D. W. Strickler. Symposium on Nucleation and Crystallization in Glasses and Melts (1962), pp. 55-62.

21. V. I. Shelyubskii and N. M. Vaisfel'd. Steklo i Keramika, No. 5:23 (1960).

22. F. Oberlie. Naturwissenschaften 43(10):224 (1956).

23. W. Vogel and K. Gerth. Glastech. Ber. 31(1):15-28 (1958).

24. W. Vogel and K. Gerth. Silikat Tech. 9(11):495-501 (1958).

25. W. Skatulla, W. Vogel, and H. Wessel. Silikat Tech. 9(2):51-62 (1958).

26. W. Vogel. Silikat Tech. 10(5):241-250 (1959).

27. L. C. Hoffman and W. O. Statton. Nature 176(4481):561-562 (1955).

28. E. A. Porai-Koshits. Glastech. Ber. 32(11):450-459 (1959).

29. W. Vogel. Z. Chem. 3(7):271-272 (1963).

30. N. M. Vaisfel'd and V. I. Shelyubskii. in: The Glassy State, Izd. Akad. Nauk SSSR (1963), p. 41. [English translation: The Structure of Glass, Consultants Bureau, New York (1964) p. 37.]

31. N. V. Solomin, V. I. Shelyubskii, and N. M. Vaisfel'd. Dokl. Akad. Nauk SSSR 140(5) (1961).

32. E. S. Sorkin and N. M. Vaisfel'd. Dokl. Akad. Nauk SSSR 151(3) (1963).

33. V. I. Shelyubskii and N. M. Vaisfel'd. Zh. Fiz. Khim. 35(11):2652 (1961).

34. I. I. Kitaigorodskii and R. Ya. Khodakovskaya. in: The Glassy State, Part 1, Izd. Akad. Nauk SSSR (1963), p. 31. [English translation: The Structure of Glass, Consultants Bureau, New York (1964) p. 27.]

35. S. I. Sil'vestrovich and É. M. Rabinovich. Zh. Vses. Khim. Obshchestva im. D. I. Mendeleeva 5(2):186 (1960).

36. S. I. Sil'vestrovich and É. M. Rabinovich. Tr. Mosk. Khim. Tekhn. Inst. im. D. I. Mendeleeva, No. 37 (1962), p. 75.

37. É. M. Rabinovich. Dokl. Akad. Nauk SSSR 138(1):159 (1961).

38. N. A. Toropov. in: The Glassy State, Part 1, Izd. Akad. Nauk SSSR (1963), p. 5. [English translation: The Structure of Glass, Vol. 3, Consultants Bureau, New York (1964) p. 5.]

39. É. M. Rabinovich. in: The Glassy State, Part 1, Izd. Akad. Nauk SSSR (1963), p. 24. [English translation: The Structure of Glass, Vol. 3, Consultants Bureau, New York (1964) p. 21.]

40. W. A. Weyl and E. C. Marboe. Glass Ind. 41(8-12):1960; 42(1-4) (1961).

41. F. P. Glasser, I. Warshaw, and R. Roy. Phys. Chem. Glasses 1(2):39 (1960).

CHAPTER III

Survey of the Literature on the System $Li_2O-Al_2O_3-SiO_2$

In studies of the crystallization kinetics of the lithium aluminosilicate glass which is the subject of the present investigation, one needs to refer continually to data on the corresponding binary and ternary systems. A survey of the literature on the subject is given below.

The components of the ternary system $Li_2O - Al_2O_3 - SiO_2$ form three binary systems, $Li_2O - SiO_2$, $Al_2O_3 - SiO_2$, and $Li_2O - Al_2O_3$. The first two systems have been studied in considerable detail, but the third has been studied very little.

The System $Li_2O - SiO_2$. This system was studied most fully by Kracek [1,2]. The phase equilibrium diagram obtained by Kracek is given in Fig. 14.

The system forms three compounds $Li_2O \cdot 2SiO_2$, $Li_2O \cdot SiO_2$, and $2Li_2O \cdot SiO_2$; the first and third melt incongruently. Lithium disilicate exists in two forms, with a transition temperature at 939°C.

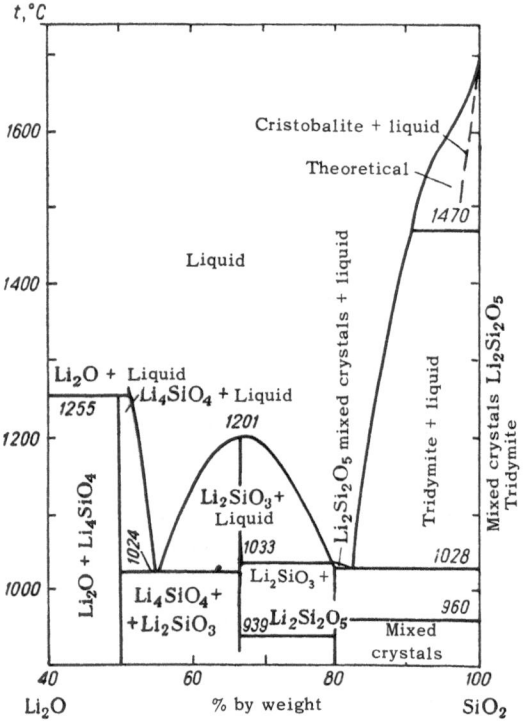

FIG. 14. Phase diagram of the system $Li_2O - SiO_2$, % by weight.

The diagram contains a narrow region of solid solutions formed by lithium disilicate with lithium metasilicate on the one hand and with silica on the other.

The liquidus curve of α-cristobalite in the system $Li_2O - SiO_2$ is S-shaped. This places the system in an intermediate position between alkali silicate and alkaline-earth silicate systems. The S-shaped curve indicates intensive separation in melts of compositions lying under this curve. The occurrence of separation is confirmed by the fact that glasses with more than 80% of silica by weight are always opalescent. The opalescence is caused by minute particles of tridymite which separates out the glass; this is confirmed by X-ray structural analysis. The glass becomes richer in lithium oxide.

Glass formation in the system $Li_2O - SiO_2$ has been studied by a number of investigators [3-5]. Blair and Urnes [5] discuss the problem theoretically; according to their data, the region of glass formation lies in the range of 27-37 mole % of lithium oxide.

The System $Al_2O_3 - SiO_2$. Studies of this system are reported in [6-9]. According to Bowen and Greig [6], mullite $(3Al_2O_3 \cdot 2SiO_2)$ melts incongruently. Subsequently Toropov and Galakhov [7] demonstrated that it melts congruently. Their phase diagram for the system $Al_2O_3 - SiO_2$ is given in Fig. 15.

The System $Li_2O - Al_2O_3$. Several short communications are concerned with this system. Two compounds, lithium aluminate $(Li_2O \cdot Al_2O_3)$ and the compound $Li_2O \cdot 5Al_2O_3$ were studied in [11-12]; it was initially thought that the latter compound is close to γ-alumina in structure [11],

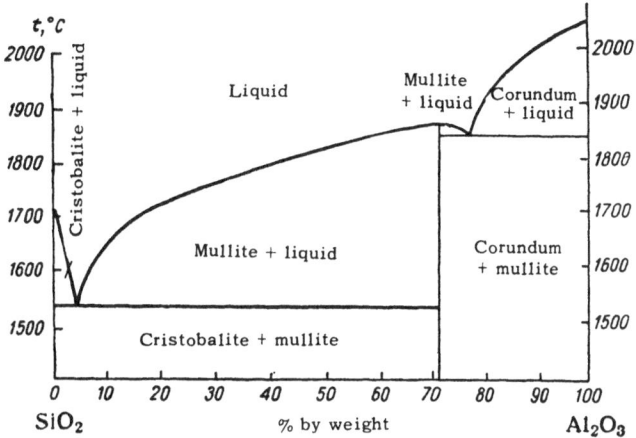

FIG. 15. Phase diagram of the system $Al_2O_3-SiO_2$, % by weight [7].

but a more detailed study showed it to have a structure of the spinel type [12]. Refinement of the phase diagram of the $Al_2O_3 - SiO_2$ system in the mullite region was published by N. A. Toropov and F. Ya. Galakhov [Izv. Akad. Nauk SSSR, Otd. Khim. Nauk. 1:8-15 (1958)].

Galakhov's paper [13] contains information on the high-alumina portion of this system. The existence of a eutectic (see Fig. 16) is noted between the crystallization fields of lithium aluminate and γ-alumina at 1760°C and a lithium oxide concentration of 14% by weight.* On the

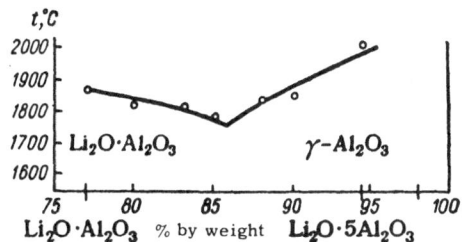

FIG. 16. Results of determinations of melting points and primary phases in the high-alumina region of the system $Li_2O-Al_2O_3$.

basis of the data of Barlett and Kordes [11-12] and also because of the similar properties of lithium aluminate and γ-alumina, Galakhov suggested that a solid solution is formed by these two compounds. At lithium oxide concentrations of up to 1%, corundum separates out as the primary phase. The phase relationships were not investigated in detail.

With reference to the ternary system $Li_2O - Al_2O_3 - SiO_2$, it must be first pointed out that three lithium aluminosilicates occur in nature: the minerals eucryptite ($Li_2O \cdot Al_2O_3 \cdot 2SiO_2$), spodumene ($Li_2O \cdot Al_2O_3 \cdot 4SiO_2$), and petalite ($Li_2O \cdot Al_2O_3 \cdot 8SiO_2$).

The first fairly detailed investigation of the ternary system $Li_2O - Al_2O_3 - SiO_2$ was carried out by Hatch [14]. As all the above-mentioned native lithium aluminosilicates lie on the silica-lithium aluminate line, Hatch chose this section for his investigation. His diagram for this section is given in Fig. 17.

With silica concentrations from 100 to 64% (the spodumene composition), the system is binary in character and separates into a silica field and a field of a solid solution of β-spodumene with silica with a eutectic point at 84.5% SiO_2 and a melting point of 1356°C. Solid solutions of the β-spodumene type extend up to approximately 76% of silica, and at higher silica contents, tridymite crystallizes out as the second phase. Compositions corresponding to petalite (78.5% SiO_2) or "lithium orthoclase" (73.2% SiO_2) are unstable at the liquidus. Only spodumene has a definite melting point, 1423°C. Spodumene ($SiO_2 \cdot Al_2O_3 \cdot 4SiO_2$) exists in two polymorphic forms: the low-temperature α form and the high-temperature β form, with a transition temperature in the 720-950°C region.

In the region of 64.6-47.7% of silica (eucryptite composition), the system behaves like a binary system above the solidus line and separates into fields of β-spodumene solid solution and a solid solution of β-eucryptite, with a peritactic point at 57.3% SiO_2. In both fields the solutions exhibit an increasing degree of dissociation to form an unknown fibrous product, resembling

*Except when stated otherwise, all concentrations are expressed in this chapter as % by weight.

21

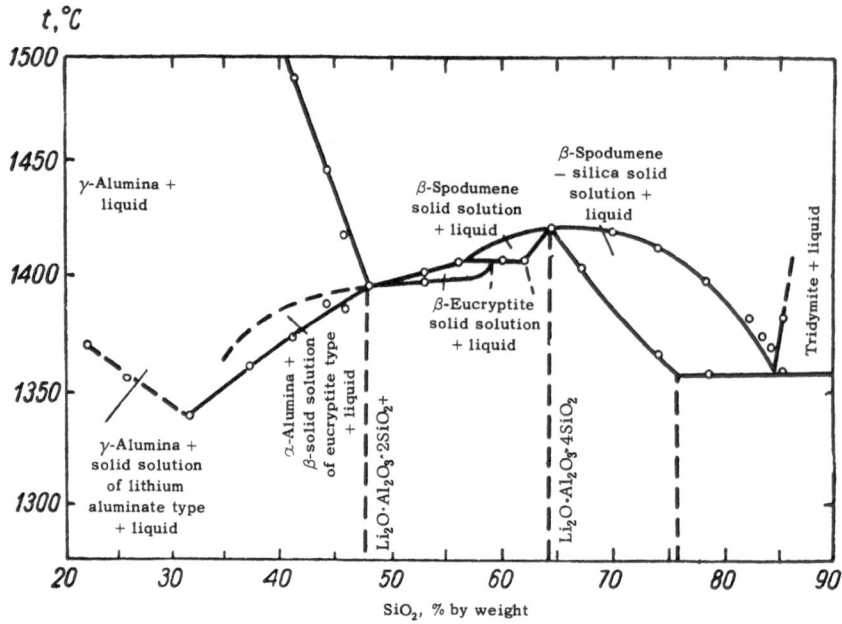

FIG. 17. Phase diagram of the pseudobinary partial system $Li_2O \cdot Al_2O_3 - SiO_2$, % by weight.

mullite, with decrease of silica content. However, according to Hatch, this is neither mullite nor a phase related to it.

At temperatures 10-80° below the solidus, a small amount of mullite is formed in solid solutions of β-spodumene, and a somewhat larger amount of mullite or α-alumina in solid solutions of β-eucryptite.

Eucryptite also exists in two forms: the low-temperature form(α), and the low-temperature form (β). The transition temperature was not determined. The β form of eucryptite is unstable and decomposes below the liquidus (1397°C) into γ-alumina and a β-eucryptite-like substance having somewhat lower optical constants than eucryptite:

Eucryptite	Eucryptite-like substance
ω 1.531	1.528
$\in$ 1.523	1.521

The system loses its binary character entirely at silica contents from 47.7 to less than 22%. The crystallization products in the composition range of 47.7-31.3% silica are γ-alumina and β-eucryptite solid solutions; at lower silica contents, they are γ-alumina and a solid solution of lithium aluminate.

The author's attempts to obtain the α forms of lithium aluminosilicates by hydrothermal synthesis were unsuccessful.

Roy et al. [15] also investigated the join $Li_2O \cdot Al_2O_3 - SiO_2$ of the diagram. Their results are given in Fig. 18. The authors confirm the existence of two crystallization fields of solid solutions of β-spodumene and β-eucryptite with silica. However, crystallization in a narrow region near a point corresponding to the 1:1:3 composition (57.7% of silica) does not produce crystals with different optical characteristics, as should be the case in a system with solid solutions, and either only crystals with a positive optical sign or only crystals with a negative optical sign are found. The authors attribute this result either to the occurrence of a small region of immiscibility, indicated by a dotted circle in the diagram of Fig. 18, or to the fact that the liquidus curves coincide at this point. The explanation for the formation of crystals with different optical signs is that the high-temperature (β) form of spodumene may change during crystallization into some form stable at lower temperatures, with negative optical characteristics. The second explanation appears to be the more probable. The authors of [15] did not succeed in obtaining an optically negative form of spodumene by low-temperature treatment of a melt of the stoichiometric composition. However, they observed that when native spodumene and petalite are melted in alumina crucibles, i.e., with excess alumina, optically negative crystals are obtained; experiments in platinum crucibles yield optically positive crystals.

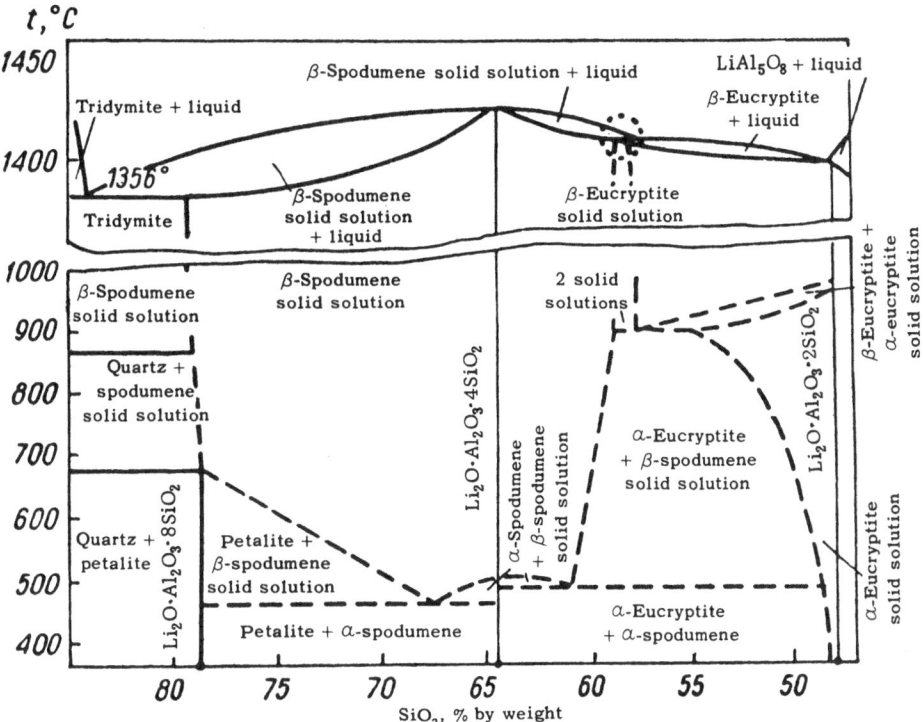

FIG. 18. Phase equilibrium diagram for the eucryptite $Li_2O \cdot Al_2O_3 \cdot 2SiO_2-SiO_2$ series, % by weight: the top part of the diagram (high-temperature region) is after Hatch, with minor modifications; the lower part is based on the results of hydrothermal experiments.

The authors investigated crystallization products formed in the subsolidus region under hydrothermal conditions. The results (indicated by dash lines, since they are not sufficiently reliable) are shown in the lower part of the diagram in Fig. 18.

Hummel [16] carried out an X-ray diffraction study of specimens with compositions lying on the silica-lithium aluminate line. The presence of a β-spodumene structure was found; no traces of the optically negative spodumene reported by Roy [15] could be detected. In compositions with high silica contents, cristobalite is found together with solid solutions of β-spodumene. In the composition $LiO_2 \cdot Al_2O_3 \cdot 12SiO_2$ quartz is also found, while cristobalite disappears after prolonged heat treatment at 1000°C.

In studies of the silica-lithium aluminate system much attention is devoted to searches for the compound $Li_2O \cdot Al_2O_3 \cdot 12SiO_2$. Saalfeld [17,18] studied the X-ray diffraction patterns of petalite and spodumene heat treated at 900-1000°C and concluded that such a compound exists. The dissociation of petalite is represented by Saalfeld as follows:

$$Li_2O \cdot Al_2O_3 \cdot 8SiO_2 \longrightarrow Li_2O \cdot Al_2O_3 \cdot 6SiO_2 + 2SiO_2$$

The crystals partly vitrify (which may indicate the formation of amorphous SiO_2).

In 1959 Roy published a paper of great significance for interpretation of the effects occurring during crystallization of compositions lying along the line $SiO_2 - Li_2O \cdot Al_2O_3$ [19]. He pointed out that silica with somewhat higher values of the cell constants is found in the crystallization products of lithium silicate glasses.

In earlier publications, Roy [15] and Heinglein [20] showed that a product with an enlarged unit cell, and β-eucryptite in some cases, may separate out from the spodumene composition.

Roy found a new form of silica (which he proposed to name silica-O in honor of Osborn) in these crystallization products by means of X-ray diffraction analysis.

In 1954 Keat discovered yet another new form of silica-keatite or silica-K [21-22]. It was shown by Shropshire [23] that the structure of β-spodumene is analogous to that of keatite. Analogy of structure leads to analogy of properties: the coefficient of thermal expansion of keatite in the 20-300°C range is small and negative, analogously to that of spodumene. There is also a far-reaching analogy between the formation of solid solutions of β-eucryptite with silica-O and of β-spodumene and silica-K.

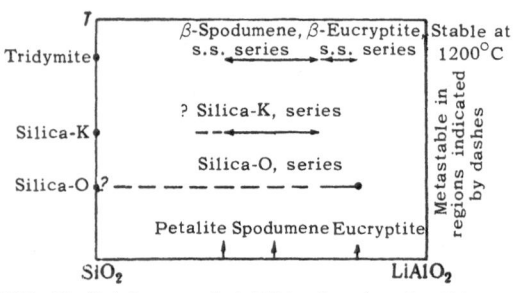

FIG. 19. Existence and stability of series of solid solutions in the system Li$_2$O · Al$_2$O$_3$–SiO$_2$.

The conventional diagram plotted by Roy [19] is shown in Fig. 19. The absissa gives compositions along the join Li$_2$O · Al$_2$O$_3$ – SiO$_2$, while the ordinate (not to scale) indicates the crystallization temperature.

The diagram shows that normal solid solutions of β-spodumene and β-eucryptite are obtained at high crystallization temperatures; at lower temperatures solid solutions of silica-K and β-spodumene crystallize out; stable formations are obtained from the composition Li$_2$O · Al$_2$O$_3$ · 8SiO$_2$ to Li$_2$O · Al$_2$O$_3$ · 3SiO$_2$. The high-silica members of the series of solid solutions of β-spodumene and silica-K are not obtainable, and this region is therefore indicated by a dash line. At even lower temperatures, in the composition range from silica to Li$_2$O · Al$_2$O$_3$ · 2SiO$_2$, solid solutions of silica-O and β-eucryptite (isostructural with silica-O) crystallize out. It was shown that when lithium aluminosilicate gels, over a wide range of compositions, are cautiously heated from 750 to 900°C, solid solutions of silica-O separate out.

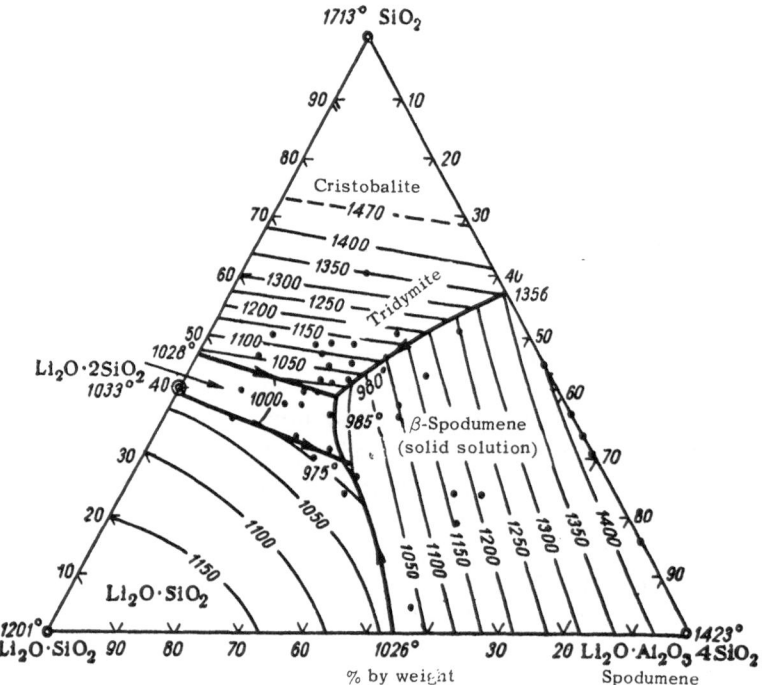

FIG. 20. The partial ternary system spodumene — lithium metasilicate — silica (compositions studied are indicated by points); % by weight.

Crystallization of new forms of silica is also observed in compositions other than those along the join SiO$_2$ – Li$_2$O · Al$_2$O$_3$.

Rindone [24] investigated crystallization of lithium disilicate nucleated by platinum and found that with relatively high platinum contents silica-O crystallizes out of the glass.

Roy and Osborn [25] studied the system lithium metasilicate-spodumene-silica. The phase diagram is shown in Fig. 20. Roy and Osborn found two eutectic points in the system; the first, for lithium disilicate-tridymite and a solid solution of β-spodumene with silica (33% SiO$_2$) is at 980°C. The second is at 975°C, at the intersection of the fields of lithium metasilicate and disilicate and solid solutions of silica in spodumene (24% SiO$_2$).

Photosensitive glasses are situated in this region of the ternary system; therefore, the results of the investigation cited are especially interesting in relation to the production of such glasses.

Galakhov [13] studied compositions in the high-alumina region of the ternary system. His results are presented in Fig. 21. He used the data he obtained earlier [10], according to which mullite (3Al$_2$O$_3$ · 2SiO$_2$) melts congruently.

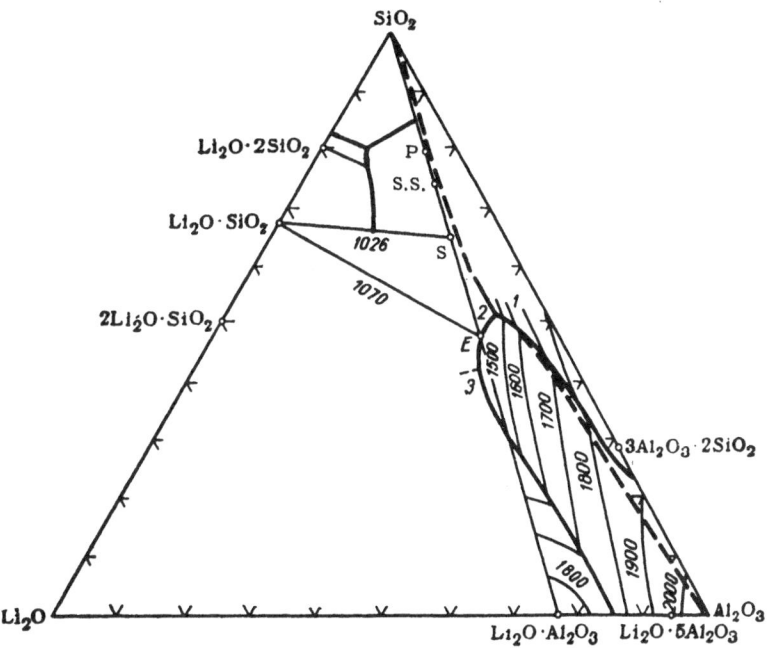

FIG. 21. Phase diagram of the system Li$_2$O–Al$_2$O$_3$–SiO$_2$ (boundaries in the alumina region): P) Li$_2$O·Al$_2$O$_3$·8SiO$_2$; S. S.) Li$_2$O · Al$_2$O$_3$ · 6SiO$_2$; S) Li$_2$O · Al$_2$O$_3$ · 4SiO$_2$; E) Li$_2$O · Al$_2$O$_3$ · 2SiO$_2$.

Figure 21 shows that the corundum field occupies a very narrow region, one of its boundaries is indicated by a dash line. The phase boundary between the γ-alumina and lithium aluminate fields crosses the silica-lithium aluminate line twice. Phase transitions occur at points 1, 2, and 3; their presumed nature is indicated in Table 2.

TABLE 2
Invariant Points in the Investigated Region
of the System Li$_2$O – Al$_2$O$_3$ – SiO$_2$

Points	Phases	Composition, % by weight			Temperature, °C
		Li$_2$O	Al$_2$O$_3$	SiO$_2$	
1	α-Al$_2$O$_3$ + γ-Al$_2$O$_3$ + 3Al$_2$O$_3$·2SiO$_2$	7	43	50	1500
2	γ-Al$_2$O$_3$+s.s. β—Li$_2$O·Al$_2$O$_3$·2SiO$_2$+3Al$_2$O$_3$·2SiO$_2$	8	41	51	1490
3	γ-Al$_2$O$_3$+s.s.Li$_2$O·Al$_2$O$_3$+s.s.β-Li$_2$O·Al$_2$O$_3$·2SiO$_2$	15	43	42	1400

Eppeler [26] studied the course of crystallization in the region to the left of the Li$_2$O · Al$_2$O$_3$ – SiO$_2$ line. Another triple point was found, at 62.5% SiO$_2$, 17% Al$_2$O$_3$, and 20.5% Li$_2$O. Simultaneous crystallization of lithium metasilicate, solid solution of β-spodumene, and solid solution of β-eucryptite occurs at this point. Silica-O often crystallizes as a metastable phase when the glasses are crystallized at relatively low temperatures. Not only the metasilicate but also a solid solution of silica-O in β-eucryptite separates out during crystallization of the glass in the lithium metasilicate field at 600°C. At 700°C the region of simultaneous presence of lithium metasilicate and solid solution of β-spodumene occupies a large area in the diagram. Formation of solid solutions of β-spodumene becomes appreciable at 800°C, and at 900°C they predominate in the crystalline mass.

Several investigations are concerned with the structure of various forms of lithium aluminosilicates and the synthesis of the corresponding melts.

Winkler [27] describes the synthesis of β-eucryptite from a $NaAlSiO_4 - LiF$ melt; the crystallization conditions and optimum composition for obtaining single crystals of β-eucryptite are given. The optical properties of β-eucryptite are compared with those of nepheline ($NaAlSiO_4$), kaliophilite, and kalsilite ($KAlSiO_4$), which are of similar structure. Winkler gives an approximate description of the β-eucryptite structure and suggests for the first time that it is similar to the structure of high-temperature quartz. Isomorphous replacement of $[SiO_4]$ by $[Al^-O_4]$ Li^{+1} in the β-eucryptite structure results in doubling of the unit cell along the C axis. $[SiO_2]$ and $[AlO_4]$ groups alternate. The charges are balanced as shown below:

$$
\begin{array}{c}
\overset{\displaystyle |}{O} \qquad \overset{\displaystyle \nearrow Li \diagdown}{\underset{\displaystyle}{}} \overset{\displaystyle |}{O} \\
| \qquad \qquad | \\
-O-\underset{\displaystyle |}{\overset{4+}{Si}}-O\underset{\tfrac{3}{4}}{-}\underset{\displaystyle |}{\overset{3+}{Al}}-O- \\
\overset{\displaystyle |}{O} \qquad \overset{\displaystyle |}{O} \\
| \qquad \qquad |
\end{array}
$$

i.e., $\tfrac{1}{4}$ of the valence of an oxygen atom is balanced by a lithium atom in fourfold coordination. As in high-temperature quartz, the chains of $[SiO_4]$ and $[AlO_4]$ tetrahedrons form spirals, slightly distorted by lithium ions.

In comparison with the structure of high-temperature quartz, the structure of β-eucryptite is somewhat weakened by isomorphous replacement of silicon-oxygen $[SiO_4]$ aluminumoxygen $[AlO_4]$ tetrahedrons. The Li-O distance in eucryptite is approximately 2.01 A, which indicates almost complete contact between the lithium and oxygen ions. The absence of any substantial cavities in this unit confers a considerable energy advantage to the eucryptite structure; this cannot be said about the structure of other alkali aluminosilicates.

Roy et al. [15] worked on the synthesis of low-temperature forms of lithium aluminosilicates. α-Eucryptite was synthesized at 300-800°C and at pressures of 70-2000 kg/cm^2. The crystals grew fairly rapidly, and were of considerable size (3-5 mm). At 972°C ± 10° and atmospheric pressure, the substance is converted into a high-temperature form having the same properties as the β-eucryptite prepared by Winkler's method [27].

With great difficulty, Roy et al. succeeded in obtaining crystals of α-spodumene. The preparation conditions were: temperature in the pressure vessel, 450°C; pressure, 800 kg/cm^2; reaction time, about two weeks. The X-ray diffraction patterns of the substance differ somewhat from those of native spodumene. At temperatures above 500°C the synthetic α-spodumene changes into the β-form.

Petalite was synthesized at 300-600°C under pressures of 70-2300 kg/cm^2. The properties of the product corresponded to those of the native mineral. Its composition lies in the range from $Li_2O \cdot Al_2O_3 \cdot 6SiO_2$ to $Li_2O \cdot Al_2O_3 \cdot 10SiO_2$. The authors favor the formula $Li_2O \cdot Al_2O_3 \cdot 8SiO_2$ on the basis of the optical properties of the crystals.

The optical characteristics of the synthetic products and native minerals described in [14, 15, 27] are given in Table 3.

Roy et al. [15] also give X-ray structural data on the lattice parameters of all known lithium aluminosilicates in the low- and high-temperature forms, and data on optically negative β-spodumene.

Barrer and White [28] confirm the data of Roy et al. [15] on the products of hydrothermal synthesis.

The question of the optically negative form of spodumene is reexamined by Heinglein [20].

TABLE 3
Optical Properties of Crystalline Phases

	α-Eucryptite		β-Eucryptite		
	synthetic	native	Roy [15]	Winkler [27]	Hatch [14]
Characteristics:					
ϵ	1.587±0.002	1.587 (?)	1.520±0.003	1.519	1.523
ω	1.572±0.002	1.572	1.524±0.003	1.524	1.531
Type	Uniaxial positive	Uniaxial positive (?)	Uniaxial negative	Uniaxial negative	Uniaxial negative
Crystal form	Quartz-type trigonal pyramids	Hexagonal outlines		Hexagonal bipyramids	

	α-Spodumene		β-Spodumene		Petalite
	synthetic	native	Roy [15]	Hatch [14]	native and synthetic
Characteristics:					
N_g	1.75	1.68	1.522	1.523	1.516
N_p	1.75	1.65	1.516	1.518	1.504
Type	Biaxial	Biaxial positive	Uniaxial positive	Uniaxial positive	Biaxial positive
Extinction angle	30°	23–27°			Parallel
Crystal form	Needles	Pyroxene	Ditetragonal bipyramids	Ditetragonal bipyramids	Needles (synthetic)

On heating native petalite in the 900–1130°C region, he obtained an optically negative phase, which he named β'-spodumene, instead of the expected optically positive spodumene. The same phase can be obtained by fusion of native petalite, from synthetic spodumene, and from the composition $Li_2O \cdot Al_2O_3 \cdot 3SiO_2$. The author suggests that this phase is close to β-eucryptite. X-ray diffraction data on the cell dimensions of the β' form of spodumene are given in the paper. These data show that the size of the unit cell increases with increased replacement of $[SiO_4]$ by $[AlO_4]$ units.

Hummel [16] gives the following data on the coefficients of expansion of lithium aluminates and aluminosilicates:

Phase	$\alpha_{25-1000°}$
$Li_2O \cdot Al_2O_3$	$124 \cdot 10^{-7}$
$Li_2O \cdot 5Al_2O_3$	$82 \cdot 10^{-7}$
$Li_2O \cdot Al_2O_3 \cdot 10SiO_2$	$5.20 \cdot 10^{-7}$
$Li_2O \cdot Al_2O_3 \cdot 8SiO_2$	$3 \cdot 10^{-7}$
$Li_2O \cdot Al_2O_3 \cdot 6SiO_2$	$5 \cdot 10^{-7}$
$Li_2O \cdot Al_2O_3 \cdot 4SiO_2$	$9 \cdot 10^{-7}$
$Li_2O \cdot Al_2O_3 \cdot 2SiO_2$	Large, negative

Liquid phase separation is not touched upon in this chapter, since it is discussed in the next. However, since phase separation, which may occur in latent form, may be presumed to exist in glasses of the system $Li_2O - Al_2O_3 - SiO_2$, crystallization effects in this system may be complicated by formation of metastable products of phase separation. This is probably what occurs in compositions close to spodumene.

Questions relating to the structure of various forms of silica, X-ray structural data on eucryptite, spodumene, and their solid solutions, and structural relationships between series of solid solutions of the general formula $Li_x \cdot Al_x \cdot Si_{1-x}O_2$ are not examined more fully here because they are discussed in detail in Chapter 3 of the book by Toropov and Barzakovskii [29].

LITERATURE CITED

1. F. C. Kracek. J. Am. Chem. Soc. 52:143 (1930).
2. F. C. Kracek. J. Phys. Chem. 34:2641 (1930).

3. H. W. Rauch, C. H. Commons, and H. H. Blau. J. Am. Ceram. Soc. 42(3):113 (1959).

4. S. K. Dubrovo. Zh. Prikl. Khim. 32(4):742 (1959).

5. G. E. Blair and S. Urnes. Glastech. Ber. 34:H-8, 391 (1961).

6. N. L. Bowen and I. W. Greig. J. Am. Ceram. Soc. 7(4):238 (1924).

7. N. A. Toropov and F. Ya. Galakhov. Dokl. Akad. Nauk SSSR 78(2):299 (1951).

8. N. A. Toropov and F. Ya. Galakhov. Izv. Akad. Nauk SSSR, Otd. Khim. Nauk 1:8-15 (1958).

9. P. P. Budnikov, S. R. Tresvyatskii, and V. I. Pushanovskii. Dokl. Akad. Nauk SSSR 98(2): 781 (1953).

10. F. Ya. Galakhov. Izv. Akad. Nauk SSSR, Otd. Khim. Nauk (1957), p. 525.

11. H. B. Barlett. J. Am. Ceram. Soc. 15:361 (1932).

12. E. Kordes. Z. Krist. 91:3 (1935).

13. F. Ya. Galakhov. Izv. Akad. Nauk SSSR, Otd. Khim. Nauk (1959), p. 577.

14. H. A. Hatch. Am. Mineralogist 28:471 (1943).

15. R. Roy, D. M. Roy, and E. F. Osborn. J. Am. Ceram. Soc. 33:152 (1950).

16. F. A. Hummel. J. Am. Ceram. Soc. 34:235 (1951).

17. H. Saalfeld. Z. Krist. 115(5/6):420 (1961).

18. H. Saalfeld. Ber. Deut. Keram. Ges. 38(7):281 (1961).

19. R. Roy. Z. Krist. 111:185 (1959).

20. E. Heinglein. Fortschr. Mineral. 34:40 (1956).

21. R. B. Sosman. Science 119:738 (1954).

22. P. P. Keat. Science 120:328 (1954).

23. I. Shropshire, P. P. Keat, P. A. Vaughan. Z. Krist. 112:409 (1959).

24. G. E. Rindone. J. Am. Ceram. Soc. 45(1):7 (1962).

25. R. Roy and E. F. Osborn. J. Am. Chem. Soc. 71(6):2086 (1949).

26. R. Eppeler. J. Am. Ceram. Soc. 46(2):97 (1963).

27. H. G. Winkler. Acta Cryst. 1:27 (1948).

28. R. M. Barrer and E. A. White. J. Chem. Soc., No. 5:1267 (1951).

29. N. A. Toropov and V. P. Barzakovskii. High-Temperature Chemistry of Silicate and Other Oxide Systems, Izd. Akad. Nauk SSSR (1963).

CHAPTER IV

Methods of Investigation

The main subject of the present investigation was a glass of definite chemical composition with somewhat less lithium oxide and somewhat more aluminum oxide than in spodumene. In addition, the glass contained a small amount (approximately 3%) of components which facilitated the melting conditions. The position of this glass is indicated in Fig. 22 by a cross.

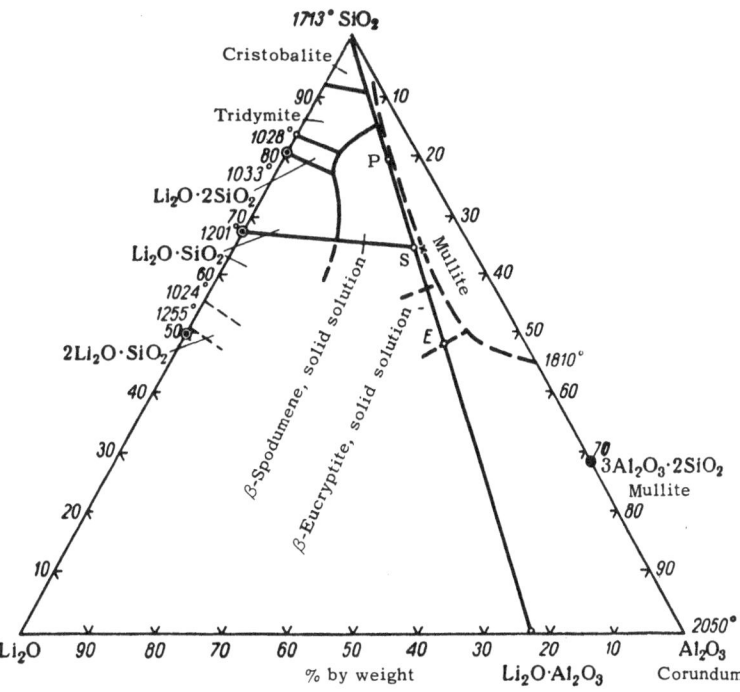

FIG. 22. Phase diagram of the system $Li_2O-Al_2O_3-SiO_2$; % by weight.
P) $Li_2O \cdot Al_2O_3 \cdot 8SiO_2$; S) $Li_2O \cdot Al_2O_3 \cdot 4SiO_2$; E) $Li_2O \cdot Al_2O_3 \cdot 2SiO_2$.

The amount of TiO_2 catalyst was varied from 0 to 11% by weight, and was 5% in most cases. The amount of titanium dioxide introduced was always on 100% of the glass. Since most of the investigations were carried out with a glass containing 5% TiO_2 as the catalyst, this glass is referred to for brevity as the original titanium glass. Additional investigations were carried out with native minerals (spodumene, etc.) and glasses corresponding in composition to such minerals as eucryptite ($Li_2O \cdot Al_2O_3 \cdot 2SiO_2$) and spodumene ($Li_2O \cdot Al_2O_3 \cdot 4SiO_2$).

Figure 22 shows that these compositions lie on a line corresponding to a 1:1 ratio of alumina to lithium oxide. The positions of eucryptite and spodumene are indicated in the diagram by circles. These compounds exist in nature as the low-temperature α forms. The high-temperature β forms always separate out during crystallization of glasses of these compositions.

The reason for this restriction in the composition of the glasses studied is that we deemed it desirable to elucidate the nature of the observed effects and to obtain a picture of the mechanism of catalyzed crystallization by application of the most diverse methods of investigation to a single object.

The original titanium glass, containing 5% TiO_2, was made in a 500-liter ceramic pot. As in the production of optical glasses, the glass was stirred by a ceramic stirrer during the melting to ensure complete homogeneity. The melted glass was cast onto a table, rolled into a sheet, and annealed. The other glasses investigated were melted in 3-liter quartz pots without stirring. These glasses were also cast into plates and annealed.

The original glass specimens were held at a series of constant temperatures in special furnaces, and in boats in a gradient furnace at temperatures from 500 to 1400°C for 24 hours.

The original glass was subjected to heat treatments in the range of 600-1000°C in steps of 10-20°. It was thus possible to study in considerable detail the nature of the structural changes taking place in the glass in the temperature range of interest. Each specimen held at a given temperature was investigated by a number of methods. This technique excluded the possibility of errors due to temperature in comparison of the results obtained by different methods and made it possible to describe fully the processes occurring in the glass.

The following methods were used for studying the catalyzed crystallization of glasses: the polythermal method, electron microscopy, X-ray structural analysis, infrared spectroscopy, differential thermal analysis, the leaching and colored indicator methods, and optical crystallography. The following physical properties were also investigated: linear thermal expansion, ΔL, in μ/cm; optical density D; density d; refractive index n_D; volume resistivity ρ_V; dielectric constant ϵ.

In the p o l y t h e r m a l m e t h o d glass specimens in ceramic boats are placed in a crystallization gradient furnace. In our experiments the boats were held in the furnace for 24 hours in temperature ranges of 500-900 or 500-1400°C [1].

The occurrence of processes in the given temperature range causing changes in the state of the glass could be assessed visually from changes in the external appearance of the glass specimens (changes of color, cracking, opalescence, surface crystallization, opacity).

In addition it was easily possible to observe in the boats the changes in the boundaries of the observed effects in relation to the composition and preliminary heat treatments of the glasses.

E l e c t r o n m i c r o s c o p e investigations were performed with the ELMI-D2 (Zeiss) electron microscope, with a resolving power of 20-30 A, at magnifications from 10,000 to 20,000.

No established methods were available when we commenced the development of technique for electron microscopic studies of transparent and crystallized glasses. Numerous techniques of specimen preparation are used in electron microscopy [2,3], but they are not all equally applicable to crystalline glass materials. The preparation technique depends to a considerable extent on the properties and nature of the material investigated; it was therefore necessary to solve anew the problem of a reliable technique for each particular case.

A series of methods for specimen preparation was tested, including the production of carbon and Parlodion replicas.

Carbon replicas were prepared by the following methods: a) with preliminary shadowing of the specimen surface by Pt-Pd alloy followed by carbon evaporation; b) by carbon evaporation followed by shadowing with Pt-Pd alloy; c) with simultaneous carbon evaporation and shadowing with Pt-Pd alloy; and d) with shadowing by chromium. Parlodion replicas were prepared by the one- and two-step methods.

Preliminary surface treatment of specimens to be studied by the replica technique is very important. Best results are obtained with freshly fractured surfaces etched with HF to reveal the structure. However, in some instances replicas obtained from freshly fractured glass surfaces without HF etching are quite satisfactory.

As the result of our investigations, the carbon replica methods with simultaneous shadowing by Pt-Pd alloy was chosen for electron microscope studies of lithium-containing crystalline glass materials. The specimen surface was previously etched with hydrofluoric acid. This technique was used in the present investigation.

The resolving power attainable with carbon replicas of moderate thickness (300-500 A) is 20-30 A.

Etching of the specimens in HF results in little distortion of the shape and size of the crystals, because the solubility of the glassy phase in the layers between the crystals is tens of hundreds of times as high as the solubility of the crystals themselves. It is therefore possible to use the form of the crystals seen in the electron micrographs for deciding whether the specimen contains one or several crystalline phases, and to draw certain conclusions concerning their composition. If the glass does not contain a crystalline phase, but has minute regions of phase separation or regions which are heterogeneous in chemical composition, the dimensions of the droplike regions observed in the electron micrographs should be regarded with great caution. The dissolution rates of the two glassy phases may be similar, so that etching produces an indistinct surface relief, with small shadows and poor contrast.

X-ray diffraction studies of catalyzed crystallization of glass were carried out with the standard URS-50I apparatus with ionization recording. The X-ray source was a tube with a copper anticathode; nickel foil was used as the filter.

Diffraction patterns of the Debye-Scherrer type were obtained both for powder specimens and by reflection of X rays from plane surfaces of glass plates. The powders were prepared from pieces of glass heated in a boat in a gradient furnace, or at constant temperature. The glass plates, about $170 \times 20 \times 5$ mm in size, were treated in the gradient furnace under the same temperature conditions. The plate surface and inner layers were investigated by reflection of X rays from successive narrow zones situated along the axis of temperature variation.

Interpretation of the results of X-ray structural analysis requires comparison of the patterns with standards. In some instances the data reported by different authors differ somewhat. These discrepancies may be caused by insufficient purity of the specimens investigated. Therefore, in addition to the literature values of the interplanar spacings of β-eucryptite and β-spodumene, we used the results of X-ray diffraction studies of crystallized glasses with the compositions of eucryptite and spodumene, and of the corresponding native minerals, for interpretation of the X-ray diffraction patterns.

Infrared reflection spectra in the 7-14 μ region were determined with the IKS-11 instrument. The specimens were polished plates. Anhydrous kerosine was used for grinding and polishing hygroscopic specimens.

After heat treatment of the specimens, both the crust layer in contact with air during crystallization and the inner regions were investigated. Nonpolarized light at an angle of about 25° was used for the reflection determinations. The precision in determinations of sharp maxima is 0.03-0.05 μ; with diffuse maxima it is lowered, sometimes considerably.

Infrared spectroscopy is one of the most sensitive methods for investigation of the structure of matter, including crystalline silicates and glasses. Infrared spectra provide a wealth of information on the structure of the crystal lattice and its most important units, such as SiO_4 and AlO_4 tetrahedrons, and AlO_6 octahedrons, and also on the structure and degree of polymerization of the silicate framework of the glass.

Some of this information may be used without additional investigations. Even the appearance of the spectra and the changes taking place in them give indications of nucleation in the course of crystallization processes, of changes produced in the material by heat treatment, of the amounts of the phases separating out, and of changes in the phase composition.

Phase composition is determined by comparison of a given infrared spectrum with a standard spectrum of the corresponding crystalline silicate. However, it is very difficult to obtain standard spectra of crystalline silicates. In particular, sufficiently reliable standard spectra have not yet been obtained for lithium aluminosilicates. Native aluminosilicates contain impurities, and their spectra vary from specimen to specimen. Mixtures of different crystalline phases may in a number of cases separate out even during crystallization of glasses of the corresponding stoichiometric compositions, so that this method cannot always be used for obtaining standard spectra.

In some cases the bands of the different silicates crystallizing out of a glass of a given composition are very close together, or their maxima may overlap. Such bands are not resolved in the spectrum. All this makes it very difficult to determine the silicates present in the original glass. Moreover, only the fundamental silicate bands lie in the infrared region easily accessible to investigation (up to 15-20 μ). The bands of a number of compounds, such as TiO_2, lie in a further region, and therefore under the usual conditions it is impossible to determine from infrared spectra whether compounds of TiO_2 are present.

Differential thermal analysis. The differential curves were recorded with the aid of a Chromel-Alumel thermocouple by means of the EPP-09 instrument with a scale range from −1 to +3 mV, at heating rates up to 33°/min. The inert substance was a powder of the original crystallized glass. In this case the zero line before appearance of a heat effect is parallel to the direction of movement of the instrument band.

The temperature at which the trace begins to deviate from this direction was taken as the temperature of the start of the effect. If a substance under investigation gives several successive heat effects, the trace between the effects has a constant deviation from the zero line. The temperature at which this deviation begins to increase was taken as the start of each successive effect.

The following characteristics were determined: the temperature t_0 of the start of the effect, the maximum temperature deviation Δt_{max} during the effect, and the time in which this deviation reaches the maximum value from zero.

The crystallization rate varies with temperature in accordance with Tammann's well-known curve. The higher the temperature, the higher is the crystallization rate, i.e., the more energy is liberated in unit time, and, therefore, the greater is the temperature deviation during the effect and the more rapidly it is completed.

Crystallization of glass becomes noticeable at the instant when the rate of heat evolution becomes comparable to the heating rate of the specimen; therefore, the higher the heating rate, the higher is the temperature at which crystallization of the specimen is detected and the heat effect begins. However, part of the heat liberated is lost by heat transfer; this decreases the magnitude of the heat effect to an extent which increases with decreasing rate of crystallization.

When the rate of heat loss equals or exceeds the rate of heat evolution, the temperature of the crystallizing specimen ceases to rise and the deviation of the differential curve becomes zero. However, this does not mean that crystallization has stopped completely. It may be detected indirectly in this temperature region by the following technique: if a glass specimen is held at constant temperature, no matter how slow the crystallization process is, the glass gradually crystallizes and its free energy decreases. If such partially crystallized glass crystallizes at higher temperatures, the corresponding heat effect is less than with the original uncrystallized glass. The extent to which the treated specimen has crystallized can be estimated from the decrease of the heat effect. Decrease of the heat effect to zero indicates that the glass has crystallized completely, and the corresponding holding time is a measure of the crystallization rate at the given temperature. Only if the heat effect given by a specimen held at constant temperature for several hundreds of hours remains unchanged can it be assumed that under these conditions the crystallization rate is virtually zero and the corresponding temperature is the boundary of the crystallization region.

The leaching method. Plates polished on all sides, 20 × 20 × 3 mm in size, were made from the glasses. They were leached in 1% HE solution in the cold for 1 hour, in 20% HCl at the boil for 1 hour, and in water at the boil for 200 hours. The weight losses of the plates during the experiments were determined. In a number of cases the products of leaching were determined quantitatively.

The colored indicator method. Crystallization of glasses was studied by the colored indicator method with the aid of Co^{2+} and Ni^{2+} ions, which are frequently used as indicators of the structure of glass, crystals, and solutions. It is known that the spectral absorption of cobalt and nickel ions alters with change of their coordination number (from four to six), while the latter, in its turn, depends on the structure of the substance [4-6]. The glass contained 0.03% CoO or 0.1% NiO. The glasses were cooled after the melting and subjected to a second heat treatment at 600-1000°C with a holding time of 24 hours. Polished specimens were prepared from the original and heat-treated glasses, and their spectral absorption was determined with the SF-4 photoelectric spectrophotometer.

Optical crystallography. The crystalline products which separated out in the glass were investigated with the MIN-8 polarizing microscope (immersion specimens and polished sections).

Linear thermal expansion, ΔL, was determined by the interference method. With specimens 7 mm thick and with the use of a portable potentiometer for temperature measurements, the precision of the determinations is of the order of $\pm 0.1\,\mu$.

The variations in the length of specimens subjected to different previous heat treatments with temperature were determined.

Optical density, D, was determined with the quartz photoelectric spectrophotometer to a precision of $\pm 2 \cdot 10^{-3}$.

Density, d, was determined by hydrostatic weighing in toluene. The precision is $1 \cdot 10^{-3}$.

Refractive index, n_D, was determined with the ITF-23 refractometer, or by Obreimov's method, to a precision of $\pm 1 \cdot 10^{-4}$.

Electrical properties. Standard methods were used for determinations of the volume resistivity ρ_V and the dielectric constant ϵ as functions of temperature: volume resistivity was measured with the E6-3 teraohmmeter, and dielectric constants in the radio frequency region with the KV-1 Q-meter. The determinations were performed in a closed cell with specimens $15 \times 15 \times 1$ mm in size, with the aid of platinum electrodes.

The furnace temperature was regulated by the EPP-09 electronic potentiometer. The temperatures of the specimens were measured with the PP potentiometer. The precision in determinations of $\log \rho_V$ was ± 0.1; ϵ was determined to within 15%.

LITERATURE CITED

1. K. G. Kumanin and E. Ya. Mukhin. Optiko-Mekhan. Prom., No. 1 (1940).
2. V. G. Luk'yanovich. Electron Microscopy in Physicochemical Investigations, Izd. Akad. Nauk SSSR (1960).
3. N. M. Vaisfel'd and V. I. Shelyubskii. in: The Glassy State, Part 1, Izd. Akad. Nauk SSSR (1963), p. 41. [English translation: The Structure of Glass, Vol. 3, Consultants Bureau, New York (1963) p. 37.]
4. T. I. Veinberg. Zh. Fiz. Khim. 36(1):81 (1962).
5. A. A. Kefeli, E. I. Galant, and N. I. Vlasova. Zh. Neorgan. Khim. 5(8):1768 (1960).
6. V. V. Vargin and T. I. Veinberg. in collection: The Glassy State, Izd. Akad. Nauk SSSR (1960), p. 372. [English translation: The Structure of Glass, Vol. 2, Consultants Bureau, New York (1960) p. 331.]

CHAPTER V

Experimental Data Obtained in Investigations of Crystallization of Lithium Aluminosilicate Glass

The original titanium glass is homogeneous, transparent, and has a yellowish color. Specimens treated at constant temperatures of 600, 630, 660, 680, 695, 705, 710, 720, 730, 740, 760, 770, 780, 790, 820, 910, and 1000° C, and also held in a boat in gradient crystallization furnace in the 550-1400°C range, differ in external appearance. After treatment at 550-620°C the color of the original glass is retained and the specimens remain transparent. Specimens treated in the 620-700°C range change color appreciably, gradually darken, and become grayish brown. They remain transparent. In the narrow range of 700-715°C the specimens crack without change of transparency or color. At 720°C cracking stops; the specimens become very turbid and whitish. This state persists to approximately 760°C. The glass then becomes transparent again, and its color changes to yellowish brown. The color gradually becomes less intense with rise of temperature to 800-810°C. On further increase of temperature the glass specimens become strongly opalescent, and at 900-1000°C and higher they become white and opaque, more like ceramics than glass in appearance.

No structural changes in the original glass after treatment at 550-620°C could be detected by any of the methods used. It is assumed that up to 620°C the glass is in its initial state or very close to it.

The results obtained by a number of workers in studies of the catalyzed crystallization of glass [1-3] and our preliminary investigations of specimens of the original titanium glass crystallized in a boat in the gradient furnace showed that in the remainder of the temperature range studied three main periods can be identified: the precrystallization region (620-700°C), the region of transparent crystallization (700-820°C), and the region of opaque crystallization (820°C and higher). The presentation of the data conforms to this subdivision.

The first results to be presented are those giving direct indications of changes in the glass structure under the influence of heat treatment (investigations by electron microscopy and X-ray diffraction, and infrared spectrum analysis). These results are examined jointly for each of the regions indicated. They are followed by the results of differential thermal analysis, chemical leaching, the colored indicator method, and investigations of changes in the physical properties of the glass as the result of heat treatment.

It should be pointed out that the temperature boundaries of the regions and individual changes within these periods depend on the thermal history of the glass [1-3]. As already stated, the glass under investigation was subjected to the usual annealing after it had been cast into a plate. It can be assumed to be in the intermediate position between the two extreme states—rapidly quenched and slowly annealed glass.

Precrystallization Region. Figure 23 shows a series of electron micrographs of freshly fractured surfaces of glass specimens subjected to heat treatments at various temperatures for 24 hours. It follows from Fig. 23,1 that heat treatment at 600°C produces no changes in the glass structure; the appearance of the surface is typical of any amorphous glass [1,4,5].

Figure 23,2 shows that when the treatment temperature is raised to 680°C small formations (0.1-0.2 μ) appear and are collected in groups of various sizes. At 695°C these formations become considerably more numerous without increasing appreciably in size (Fig. 23,3).

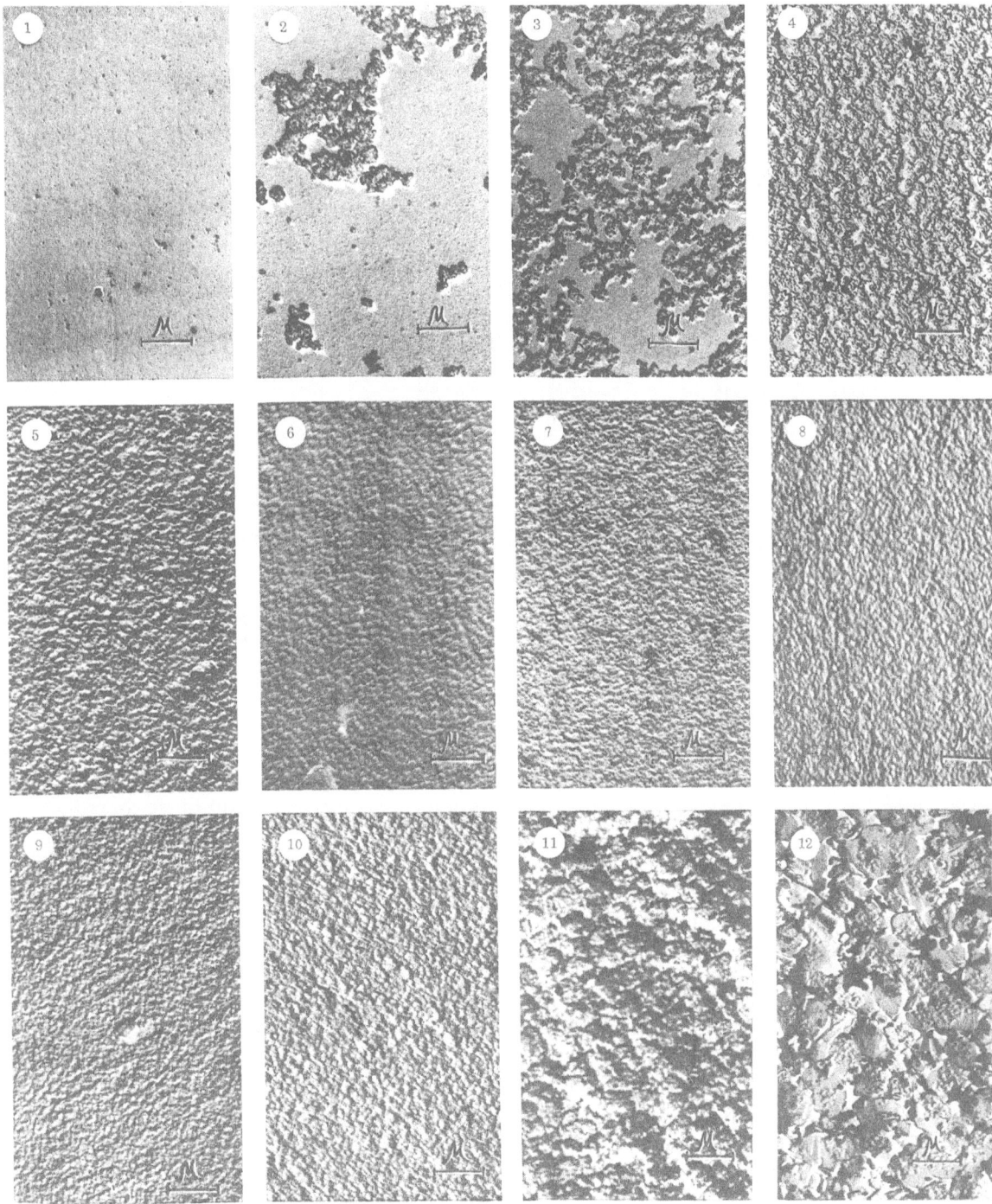

FIG. 23. Electron micrographs of glass specimens subjected to heat treatment for 24 hours at various temperatures: 1) 600°C, 2) 680°C, 3) 695°C, 4) 705°C, 5) 710°C, 6) 720°C, 7) 740°C, 8) 760°C, 9) 790°C, 10) 820°C, 11) 910°C, 12) 1000°C.

Figure 23,4 shows that at 705°C the structure of the glass becomes even more fine-grained. The grain size does not exceed 0.1 μ.

The results of X-ray diffraction investigations of the same specimens are presented in Fig. 24. The diagram shows that the X-ray diffraction patterns of specimens heated at temperatures from 600 to 680°C are of a character typical for glass (diffuse maxima).

A qualitative change appears at 695°C. An appreciable peak characteristic of crystals appears on the diffuse maximum, indicating the start of crystallization. The X-ray scattering curve is of the same character at 705°C.

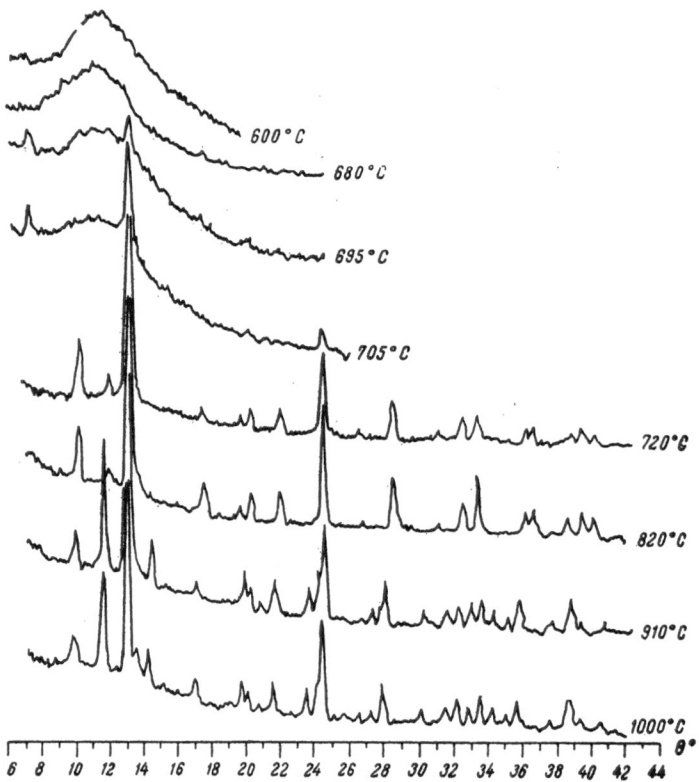

FIG. 24. X-ray diffraction diagrams of glass specimens subjected to
heat treatment at various temperatures.

Infrared reflection spectra were determined for the glass specimens treated at the same temperatures (Fig. 25). The curves for 600-695°C coincide with the spectrum of the original glass. The spectrum of the specimen heated at 705°C shows a group of small peaks, indicating the start of structural changes in the glass.

Region of Transparent Crystallization. It was noted earlier that at the start of the region of transparent crystallization the glass cracks without changing in external appearance. The cracking ceases at 720°C, when the glass becomes very turbid. The turbidity persists to about 760°C, when the glass becomes transparent again and its color changes to yellowish brown.

The electron micrographs (Fig. 23, 4-10) of glass specimens after heat treatment at 705-820°C show that the microgranular structure which appears in the glass after treatment at 705°C does not undergo any appreciable changes over the entire temperature range studied. The grain size is about 0.1 μ.

The X-ray diffraction diagrams in Fig. 24 show that the amount of crystalline phase increases sharply with rise of the temperature of heat treatment from 695 to 720°C. In the range of 720-820°C the X-ray diffraction diagrams remain substantially without change.

The infrared reflection spectrum of the specimen heated at 710°C (Fig. 25) shows that the principal maximum, which was of the serrated type at 705°C, splits into two distinct peaks at the higher temperature, and a third peak appears in the long-wave region, at 13.4 μ, indicating the separation of crystalline phases. The barely perceptible broadening of the spectrum in the region of 8.6 μ, observed at lower temperatures, becomes more distinct. The spectrum for 720°C is the same as for 710°C, except that the broadening at 8.6 μ becomes a distinct plateau. This type of spectrum persists up to 820°C.

Region of Opaque Crystallization. At temperatures above 820°C the glass specimens rapidly become turbid and then white and opaque. The transition to opaque crystallized glass occurs in a relatively narrow temperature range, the lower limits of which, known as the temperature of loss of transparency, depends on the previous heat treatment of the glass [2].

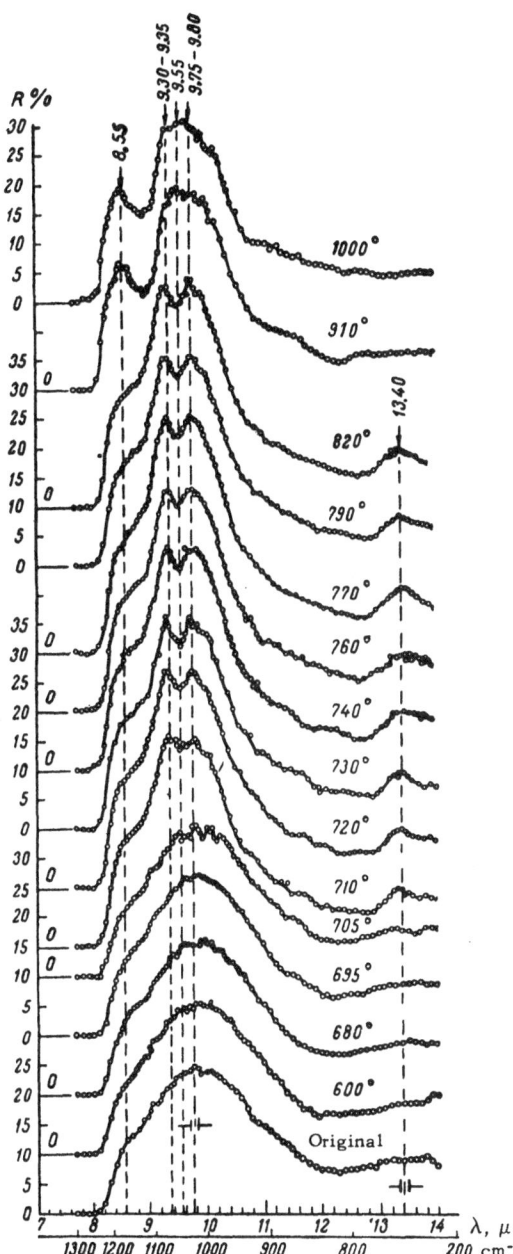

FIG. 25. Infrared reflection spectra of glasses after
heat treatment at various temperatures.

Electron micrographs (Fig. 23, 10-12) of the glass specimens show a considerable coarsening of the structure in the 820-1000°C range. Blocks 0.2-0.5μ in size, consisting of smaller formations, appear at 910°C. At 1000°C a considerable proportion of these smaller formations disappears, and larger regions (0.4-0.6μ) with noticeable faceting appear. The nature of the observed structural changes suggests that in this temperature range the fine-grained structure is transformed into larger formations (Fig. 23, 12).

The structural changes in the glass in the range of 820-1000°C and higher, described above, can be seen very clearly in the electron micrographs of specimens of the same glass after heat treatment in a boat in a gradient crystallization furnace at 740-1350°C (Fig. 26). Three types of crystals can be seen in the structure of the specimens held at 1000-1300°C. The main mass consists of large crystals, closest to hexahedrons in shape. In addition to the coarse-grained phase there is also a fine-grained phase which is generally located in the centers of the large blocks. It is similar in appearance to the formations observed in the 705-820°C range. The third phase consists of prismatic crystals. Increase of temperature from 1000 to 1300°C leads to progressive increase of the crystal size (Fig. 26, 3-7). Above 1300°C the specimen melts and its crystalline structure disappears (Fig. 26, 8).

The average size of the crystals formed during heat treatment at progressively rising temperatures was determined from this series of micrographs. The results are presented in Table 4.

X-ray structural analysis of the specimens showed that at heat-treatment temperatures above 820°C the character of the diffraction pattern changes appreciably, indicating changes in the phase composition of the crystallization products (Fig. 24). The changes of phase composition are accompanied by visible changes in the glass specimens: when heated to 820°C the glass, containing submicrocrystals, remains transparent whereas at higher temperatures it becomes opaque.

The character of the infrared reflection spectra varies analogously to the changes in the X-ray diffraction patterns. Figure 25 shows the spectra of glass specimens heated at 910 and 1000°C. They differ noticeably from the spectra obtained at lower temperatures (in the region of transparent crystallization). The plateau at 8.55 μ is replaced by a well-defined maximum. The maxima at 9.35 and 9.80 μ diminish considerably, a new maximum appears at 9.55 μ in the place of the earlier maximum, and the long-wave maximum at 13.4 μ disappears. At 1000°C only a distinct peak at 8.55 μ remains, while the remaining maxima become even more diffuse.

Thermograms in the temperature range up to 1100°C were recorded for the original glass and for specimens after heat treatment. The thermogram of the original glass is shown in Fig. 27. It may be seen that the glass exhibits two exothermic effects in this temperature range, indicating crystallization of two phases. The first effect, at about 800°C, corresponds to crys-

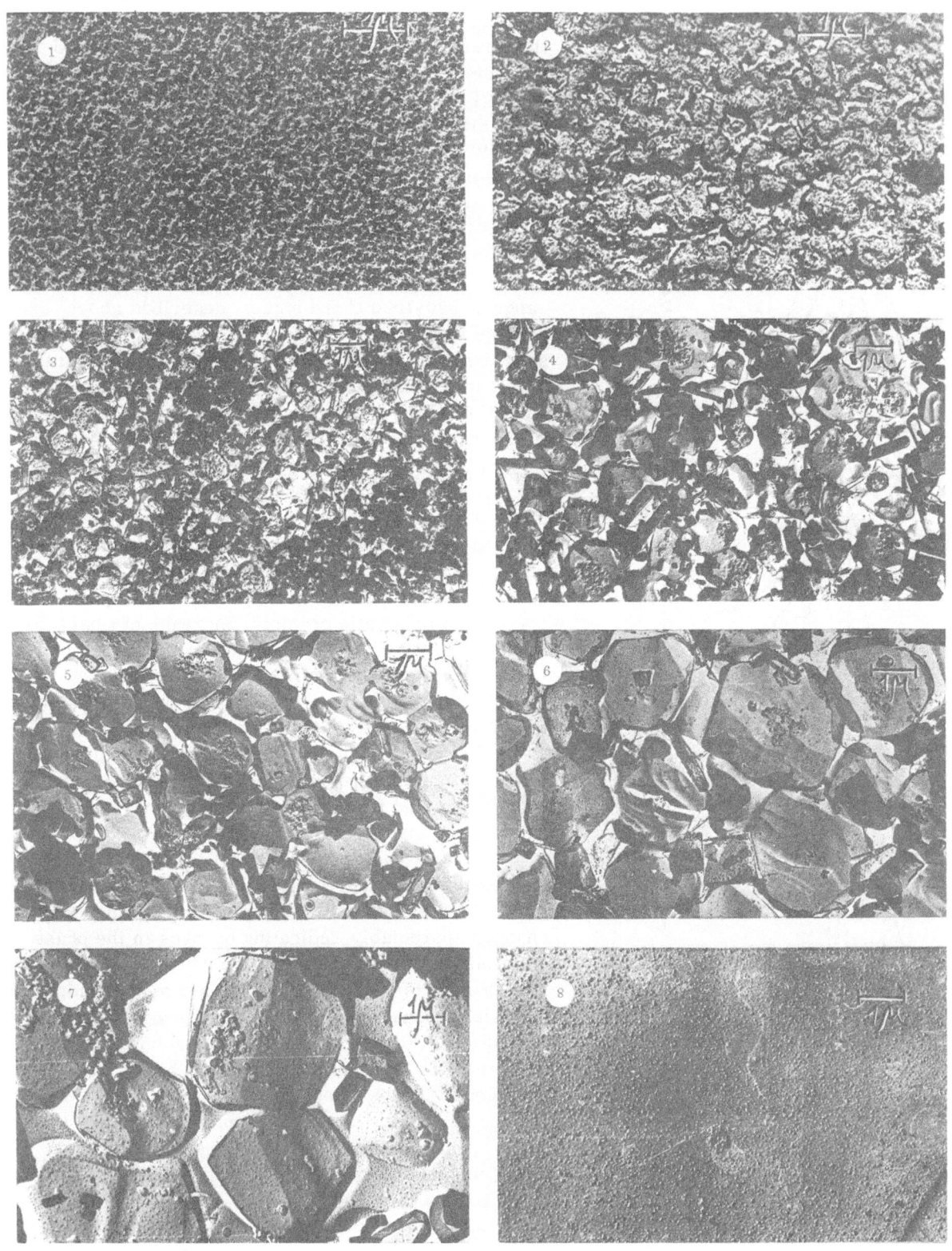

FIG. 26. Electron micrographs of glass specimens after heat treatment for 24 hours in a boat in a gradient furnace: 1) 740-820°C, 2) 820-940°C, 3) 940-1030°C, 4) 1030-1100°C, 5) 1100-1150°C, 6) 1150-1220°C, 7) 1220-1300°C, 8) 1300-1350°C.

TABLE 4

Sizes of the Crystalline Particles Formed During Heat Treatment of the Original Titanium Glass

Specimen No.	Range, °C	Average particle size, μ
1	740—820	0.12
2	820—940	0.20
3	940—1030	0.50
4	1030—1100	0.80
5	1100—1150	1.50
6	1150—1220	2.50
7	1220—1300	3.50

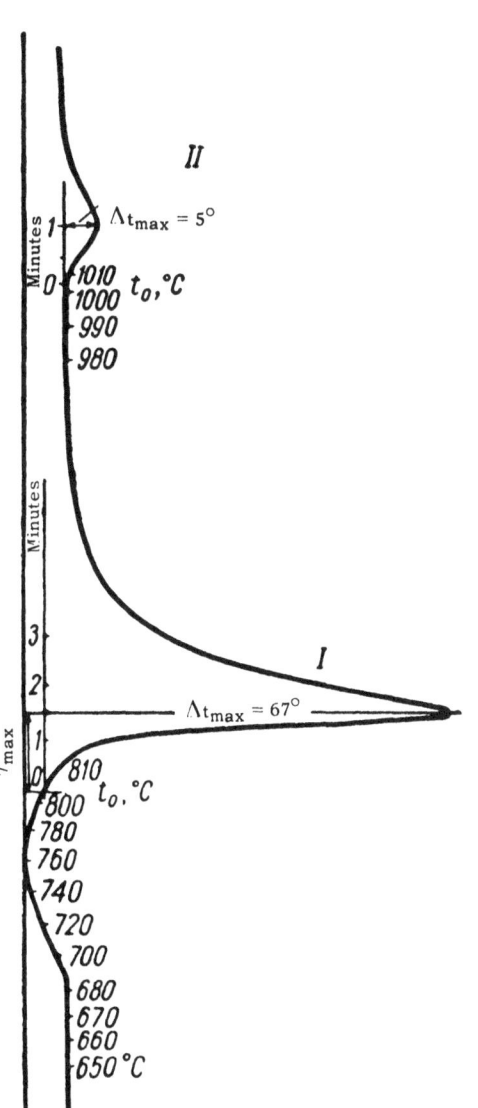

Fig. 27. Thermogram of the original glass. I and II) Regions of crystallization of the first and second phases.

tallization of the first phase to separate out of the glass. The second effect, observed at about 1000°C, indicates the presence of a second, more refractory, phase in this glass.

Table 5 gives the characteristics of the thermograms (t_0, Δt_{max}, and τ_{max}) for the original glass and for specimens after heat treatment.

The table shows that heat treatment at temperatures up to 695°C does not result in any noticeable changes of either of the heat effects. After treatment at 705°C the first effect begins to diminish, and disappears entirely after treatment at 720°C. The second heat effect remains constant up to 820°C, but treatment at 910°C causes this effect to disappear too.

These results show that the first phase crystallizes during 24 hours of heat treatment at 705-720°C, and the second at 820-910°C.

Glass specimens subjected to different heat treatments were also leached with different reagents (HF, HCl, NaOH). It was found that the leachability by the action of all these reagents decreases regularly in the sequence from the original glass not subjected to additional heat treatment, to specimens treated at 760-780°C, and then increases again; the increase is especially rapid for heat-treatment temperatures above 820°C.

The greatest variations of leachability with the temperature of heat treatment were found when 20% hydrochloric acid was used; this solution was used for more detailed investigations (Fig. 28).

Figure 28 shows that the minimum leachability (treatment temperature 760-780°C) is between a quarter and a fifth of the leachability of glass treated at 600°C. The leachability of glass held at 900°C is 40 times the minimum value; after heat treatment at 1000°C, it is 60 times the minimum.

These sharp changes of leachability undoubtedly indicate considerable structural changes in the material.

Chemical analysis of the hydrochloric acid extracts showed little variations in the compositions of extracts of the original glass and specimens treated at temperatures up to 780°C (Fig. 29). The relative proportions of the oxides in these extracts (with the exception of silica) correspond fairly closely to the composition of the original glass. The lowering of the silica content in the hydrochloric acid extracts is due to the fact that silica is only slightly soluble in hydrochloric acid and remains on the surface of the leached specimens, forming a protective film.

Thus, the processes occurring in the region of liquid phase separation and of the start of crystallization (transparent material) cause only the general decrease of leachability shown in Fig. 28, without any substantial changes in the chemical composition of the extracts.

39

TABLE 5

Characteristics of the Thermograms Recorded for the Glass Specimens at the Heating Rate of 29°/min

Treatment temperature, °C	First exothermic effect			Second exothermic effect		
	t_0, °C	Δt_{max}, °C	τ_{max}	t_0, °C	Δt_{max}, °C	τ_{max}
Original specimen	810	72	1 min 58 sec	1007	5	1 min 00 sec
To 695	809	70	1 » 35 »	1000	7	1 « 19 »
705	777	33	2 » 16 »	1006	5	1 » 57 »
710	783	23	1 » 50 »	997	7	1 » 23 »
720—820	0	0	0	995	5	1 » 34 »
910—1000	0	0	0	0	0	0

When the temperature of heat treatment is raised to 850°C and higher, and the glass becomes first opalescent and then opaque, the total leachability and the proportions of the oxides in the extracts change sharply.

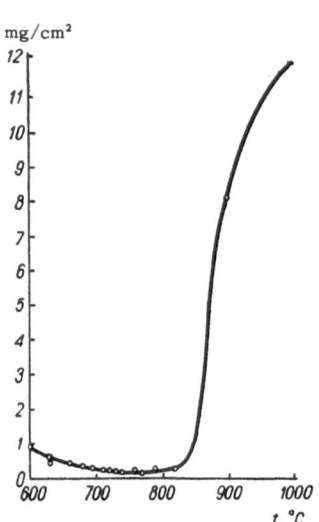

FIG. 28. Effect of heat-treatment temperature on the leachability of glass by 20% hydrochloric acid solution.

The relative silica content in the extracts falls to a very low value (approximately 5 mole %) while the lithium oxide content rises sharply. The high total leachability and the predominant contents of lithium and aluminum in the extracts indicate that a different crystalline phase, of lower cnemical stability, is formed at higher temperatures of heat treatment.

For investigations of glass crystallization by the colored indicator method, Co^{2+} and Ni^{2+} ions were used. The spectral absorption of these ions changes sharply with change of their coordination numbers from 4 to 6; the coordination numbers, in their turn, depend on the structure of the substance [6-8].

The glasses containing the colorants were melted in 3-liter pots, cast into plates, and annealed.

Figure 30 shows spectral absorption curves of glass containing Ni^{2+} not subjected to heat treatment, and after treatment for 24 hours at 655 and 725°C.

The untreated glass is yellow, and its spectral absorption indicates that nickel is present in sixfold coordination [9]. The spectral absorption of glass held at 655°C shows little difference, but the curve becomes more steep. A weak maximum at $\lambda \approx 700$ mμ and a weak minimum at $\lambda \approx 900$ mμ appear. All traces of the structure containing nickel in fourfold coordination disappear entirely from the spectral absorption curve. The glass acquires a pale green color which is due to a flat minimum at $\lambda \approx 550$ mμ.

When the glass is held at 725°C its color changes from yellow to purple and the spectral absorption changes sharply. Maxima characteristic of Ni^{2+} in fourfold coordination appear on the absorption curve [9]. This color change begins at 715°C. It should be noted that glasses of the system $Li_2O - SiO_2$, with any proportions of the components, contain Ni^{2+} in sixfold coordination only and are colored yellow. On introduction of aluminum the coordination of Ni^{2+} is still likely to become fourfold [10].

Therefore, the change in the color of the glass from yellow to purple can only be due to formation of a crystalline phase into which Ni^{2+} passes and in which the coordination number of aluminum is four.

These data show that changes in the glass structure, which are not associated with crystallization, occur at temperatures up to 700°C. Crystallization begins at 710-715°C. Spectral absorption curves of glass containing Co^{2+} are not given here, although they also show convincingly that cobalt changes from sixfold to fourfold coordination at the same temperatures of heat treatment as in glasses with Ni^{2+} [11].

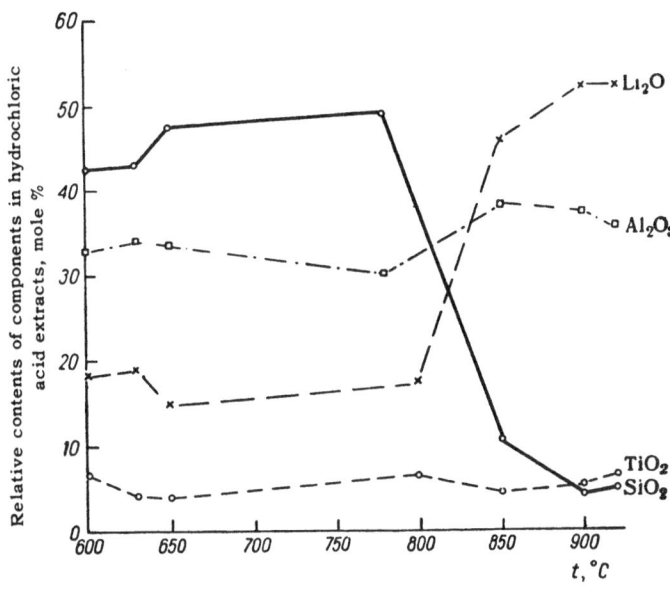

FIG. 29. Variations of the chemical composition of extracts (mole %)
with the temperature of heat treatment.

The linear expansion ΔL_1 on heating to temperatures of the order of 400–500°C was determined for the original glass and for specimens after heat treatment. The elongation curves are shown in Fig. 31.

These curves show that specimens of the original glass and glass subjected to heat treatment at temperatures up to 700°C expand so rapidly that only the regions of the curves up to 200°C could be accommodated in the diagram. It was found that up to 400°C the elongation of these specimens is a linear function of temperature, but at higher temperatures (up to 600°C) the elongation rate begins to increase somewhat. The coefficients of expansion of the original

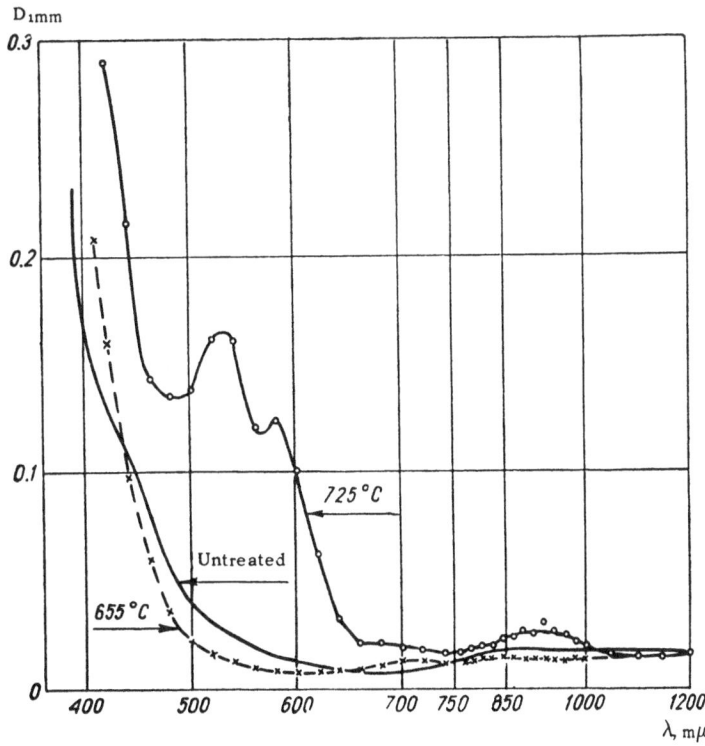

FIG. 30. Spectral absorption of glass containing Ni^{2+}, after different
heat treatments.

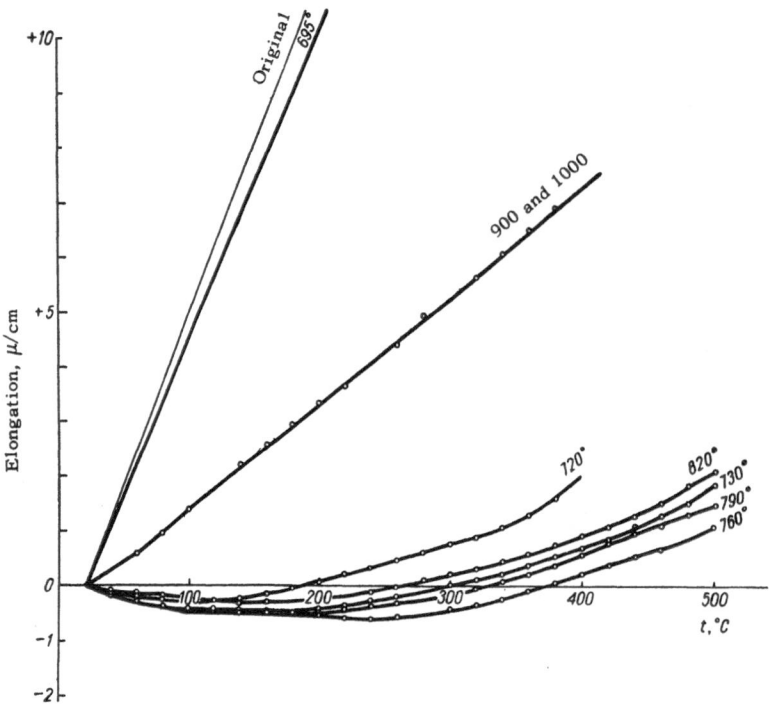

Fig. 31. Thermal expansion of glass specimens after heat treatment (the treatment temperatures are indicated on the curves).

glass and of glass after heat treatment at 695°C were 62 · 10⁻⁷ and 57 · 10⁻⁷, respectively, in the temperature range of 20-400°C; i.e., of the same order of magnitude as those of many ordinary glasses. It follows that heat treatment at temperatures up to 700°C does not produce appreciable structural changes in the glass.

The curves in Fig. 31 show that the nature of the elongation of the specimens on heating alters sharply when the temperature of heat treatment is raised from 700 to 820°C. The specimens first contract on heating and then begin to expand. The course of the elongation curves shows that the coefficient of expansion varies continuously with change of temperature.

With increase of the treatment temperature to 760°C the elongation curves shift progressively lower in relation to the abscissa, i.e., at the start of the heating the contraction is progressively greater and occurs over a progressively wider temperature range; correspondingly, the subsequent elongation up to 500°C becomes progressively less.

On further increase of temperature from 760 to 820°C the elongation curves again begin to approach the abscissa. The contraction and the temperature range in which it occurs gradually diminish.

These elongation curves for specimens subjected to heat treatment in the second temperature range (700-820°C) show that considerable structural changes occur in the glass during such treatment, mainly at 700-720°C.

Finally, the elongation curve for the specimen treated at 900°C shows that only this specimen expands on heating. The expansion is linear up to 400°C. The coefficient of linear expansion is 16 · 10⁻⁷.

The expansion of the specimen treated at 1000°C took the same course as that of the specimen treated at 900°C. Therefore two temperatures, 900 and 1000°C, are given on the corresponding curve.

It follows that heat treatment at 900°C produced further structural changes in the glass [3, 12].

Spectral curves showing variations of optical density in the wavelength range of 400-1200 mμ were recorded for all the glass specimens subjected to heat treatment at constant temperatures. The optical density at three different wavelengths (500, 600, and 700 mμ) was plotted against the treatment temperature for each of the specimens (Fig. 32). The optical density of the original glass is the same as that of the specimen treated at 600°C. Figure 32 shows that all three curves are of similar form.

Heat treatment at 600-680°C results in a gradual increase of the optical density of the specimens. In the 680-710°C range the optical density remains almost unchanged. The specimens become dark brown but remain transparent. On increase of the treatment temperature to 720°C the specimens become very turbid; the turbidity persists to 740-750°C, so that the optical density increases considerably, as is clear from the sharp rise of the curves in Fig. 32. On further increase of temperature to 790°C the turbidity diminishes considerably but does not disappear. The specimens become transparent again. The measured value of the optical density becomes somewhat lower than for specimens treated at 660-710°C, but the value is probably somewhat higher than the true optical density owing to the turbidity present. The optical density remains unchanged with increase of the treatment temperature to 820°C and the turbidity does not increase.

In Fig. 32 the points for 710 and 790°C on the optical density curves are joined by dash lines which represent the probable course of the curves if the rise due to turbidity is excluded.

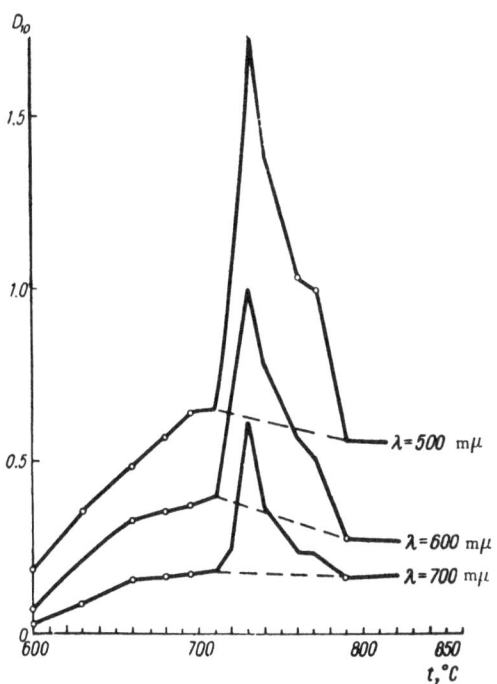

FIG. 32. Variations of optical density of glass specimens with the temperature of heat treatment, for three different wavelengths λ.

When the specimens are heated at temperatures above 820°C they again become turbid very rapidly, and above 900°C they become completely opaque. Turbidity in the 710-790°C range can be prevented by preliminary heat treatment of the glass at approximately 630°C. The turbidity which appears above 820°C is destroyed only if the glass is remelted.

Density determinations (Table 6) show that up to 630°C the density remains unchanged, within the limits of experimental error. This is followed by slight increase of density in the 630-705°C range. In a considerably narrower temperature range, 705-720°C, the density increases more rapidly. At 720-820°C the density of the glass continues to increase, and at higher temperatures it begins to decrease [3,12].

The data in Table 6 show that up to 630°C the refractive index remains the same as that of the original glass. It increases between 630 and 720°C, especially in the 705-720°C range. The increase of n_D gradually slows down between 730 and 820°C, and in the 820-870°C range, n_D tends to decrease. The glass rapidly becomes opaque at higher temperatures and determinations of refractive index become impossible [3,12].

The variations of refractive index with the temperature of heat treatment are analogous to the density variations. The specific refraction $[(n^2 - 1)/(n^2 + 2) \cdot 1/d$ was calculated for each temperature of heat treatment from the experimental values of the density d and refractive index n_D in order to reveal the causes of the change of refractive index during heat treatment.

Within the precision limits of determinations of d ($\pm 1 \cdot 10^{-3}$) and n_D ($\pm 1 \cdot 10^{-4}$), the specific refraction remained constant after heat treatment in the entire temperature range from 600 to 870°C, and was equal to the value for the original glass, or for the glass after treatment at 600°C $[(n^2 - 1)/(n^2 + 2) \cdot 1/d = 0.129]$. This shows that the change of refractive index during heat treatment is due mainly to density changes.

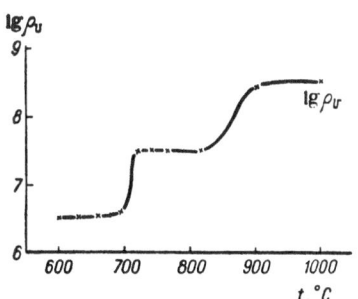

FIG. 33. Variations of volume resistivity ρ_v of glass specimens with the temperature of heat treatment.

Determinations of the volume resistivity ρ_v of specimens after heat treatment (Fig. 33) show that it varies in two temperature ranges: 700-720°C and 820-900°C. Between these ranges ρ_v remains constant [3].

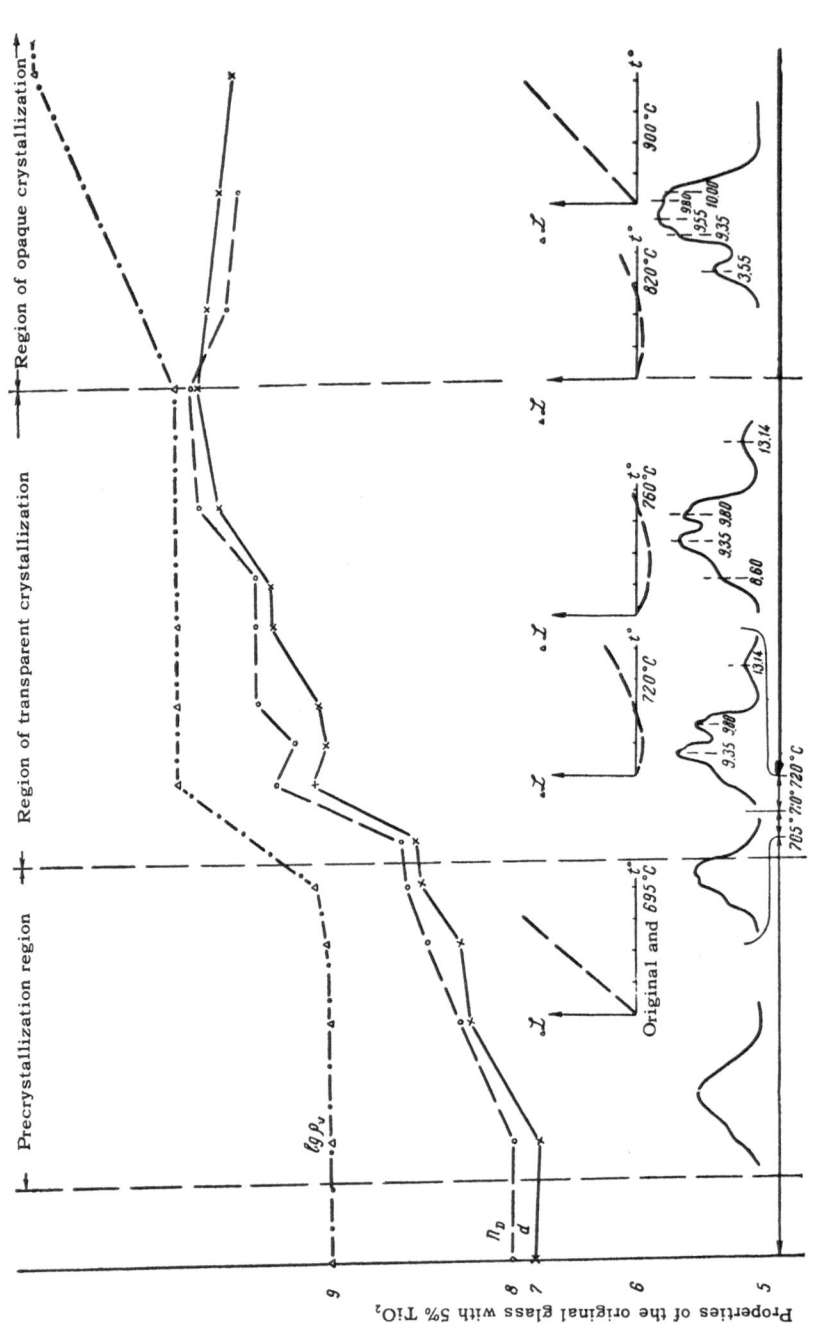

Region of opaque crystallization

Region of transparent crystallization

Precrystallization region

900°C

820°C t°

13.14

760°C t°

9.35 9.80

8.60

720°C

13.14

9.35 9.80

705° 710° 720°C

695°C t°

Original and 695°C

lg ρ_v

n_D

d

Properties of the original glass with 5% TiO₂

44

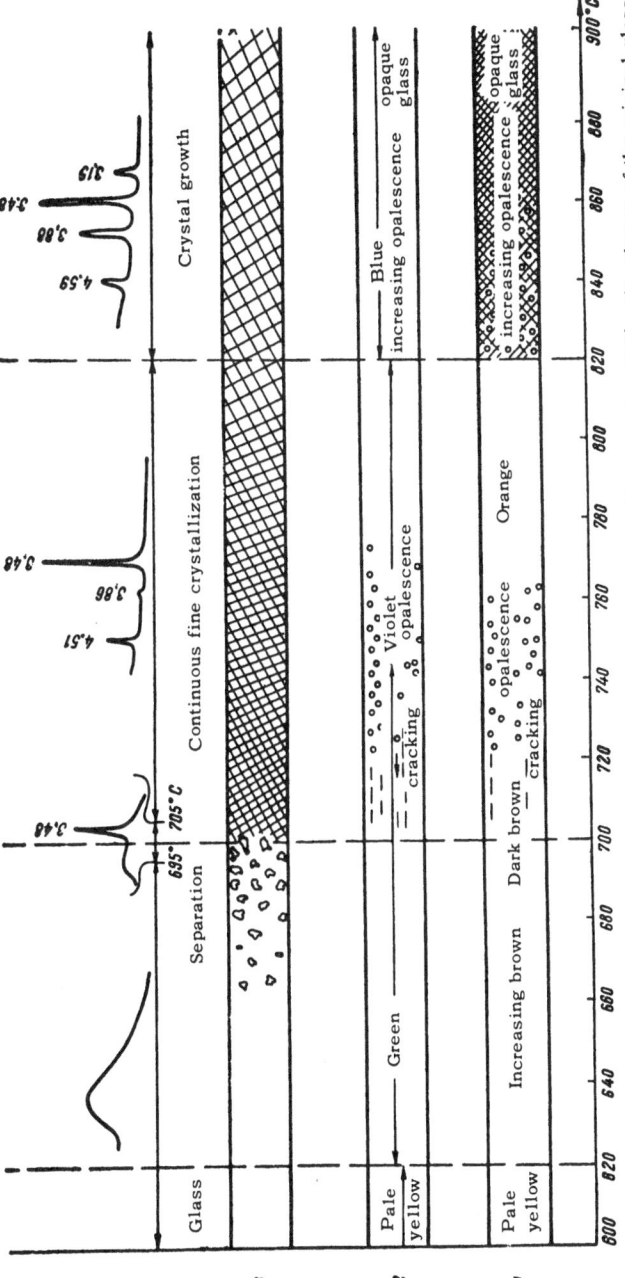

FIG. 34. Variations of the properties of the original glass with the temperature of heat treatment: 1) color change of the original glass; 2) color change of the glass containing nickel; 3) results of electron microscope studies; 4) form of X-ray diagrams; 5) form of infrared reflection spectra; 6) linear expansion, ΔL; 7) density, d; 8) refractive index, n_D; 9) logarithm of volume resistivity, $\lg \rho_V$.

45

The dielectric constants ϵ of the same specimens were determined. Plots of ϵ versus treatment temperature had inflections of the same temperatures (700-720°C and 820-900°C).

For greater clarity and convenient comparison, all the results obtained for the glass have been presented in a single diagram (Fig. 34). The results of the electron-microscopic, X-ray diffraction, and infrared spectroscopic investigations and of the elongation determinations are indicated in conventional form in Fig. 34.

All the experimental results confirm the validity of the subdivision of the temperature range studied into three regions: the precrystallization, the transparent crystallization, and the opaque crystallization region.

TABLE 6

Variations of Density and Refractive Index
with the Temperature of Heat Treatment

Temperature of heat treatment, °C	n_D	d
Original		
600	1.5400	2.447
630	1.5399	2.446
660	1.5413	2.454
680	1.5421	2.456
695	1.5426	2.459
705	1.5427	2.459
720	1.5460	2.475
730	1.5455	2.472
740	1.5465	2.473
760	1.5465	2.477
770	1.5465	2.478
790	1.5480	2.483
820	1.5482	2.487
840	1.5472	2.485
870	1.5469	2.484

LITERATURE CITED

1. I. I. Kitaigorodskii and R. Ya. Khodyakovskaya. in: The Glassy State, Part 1, Izd. Akad. Nauk SSSR (1963), p. 31. [English translation: The Structure of Glass, Vol. 3, Consultants Bureau, New York (1964) p. 27.]
2. E. V. Podushko and A. B. Kozlova. in: The Glassy State, Part 1, Izd. Akad. Nauk SSSR (1963), p. 74. [English translation: The Structure of Glass, Vol. 3, Consultants Bureau, New York (1964) p. 77.]
3. I. M. Buzhinskii, E. I. Sabaeva, and A. N. Khomyakov. in: The Glassy State, Part 1, Izd. Akad. Nauk SSSR (1963), p. 127. [English translation: The Structure of Glass, Vol. 3, Consultants Bureau, New York (1964) p. 133.]
4. V. N. Vertsner and L. V. Degteva. in: The Glassy State, Part 1, Izd. Akad. Nauk SSSR (1963), p. 83. [English translation: The Structure of Glass, Vol. 3, Consultants Bureau, New York (1964) p. 86.]
5. W. Vogel. 63rd Annual Meeting of the American Ceramic Society, Toronto, 1961.
6. T. I. Veinberg. Zh. Fiz. Khim. 36(1):81 (1962).
7. A. A. Kefeli, E. I. Galant, and N. I. Vlasova. Zh. Neorgan. Khim. 5(8):1768 (1960).
8. V. V. Vargin and T. I. Veinberg. in collection: The Glassy State, Izd. Akad. Nauk SSSR (1960), p. 372. [English translation: The Structure of Glass, Vol. 2, Consultants Bureau, New York (1960) p. 331.]
9. V. V. Vargin. Production of Colored Glass, Gizlegprom (1940).
10. H. Saalfeld. Ber. Deut. Keram. Ges. 38(7):281,286 (1961).
11. V. V. Vargin. in: The Glassy State, Part 1, Izd. Akad. Nauk SSSR (1963), p. 107. [English translation: The Structure of Glass, Vol. 3, Consultants Bureau, New York (1964) p. 114.]
12. I. D. Tykachinskii and E. S. Sorkin. in: The Glassy State, Part 1, Izd. Akad. Nauk SSSR (1963), p. 123. [English translation: The Structure of Glass, Vol. 3, Consultants Bureau, New York (1964) p. 129.]

CHAPTER VI

Influence of Titanium Dioxide on the Catalyzed Crystallization of Glass

For elucidation of the role of TiO_2 as a catalyst of volume crystallization, investigations were carried out on the original glass without TiO_2 and with various amounts of TiO_2 after heat treatment at various temperatures in boats in a gradient furnace, or in furnaces at constant temperatures.

Figure 35 shows the changes observed visually in glasses subjected to heat treatment in a gradient furnace at 500-1000°C for 24 hours. There is a similarity between the behavior of glass without TiO_2 and of glasses with 1 and 2% TiO_2. In this series of glasses crystallization begins at the surface, at temperatures in the region of 850°C, and is accompanied by cracking of the specimens. The glass specimens crystallize at treatment temperatures of approximately 900°C and higher.

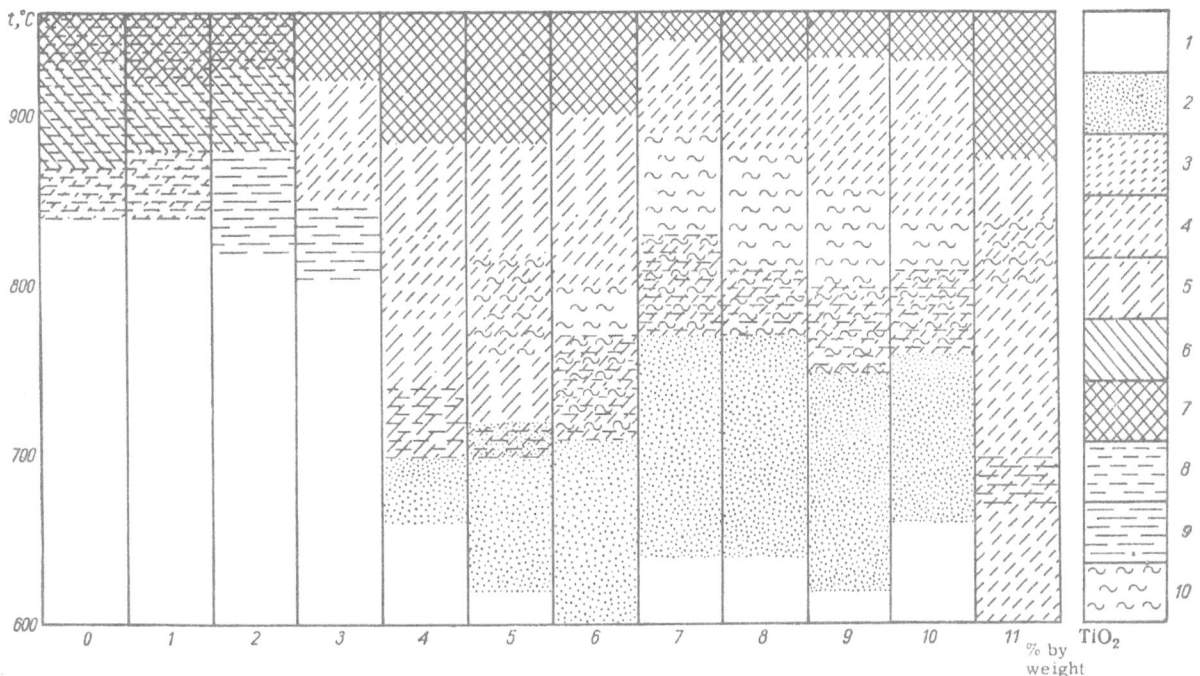

FIG. 35. Visual characteristics of glass crystallization in relation to the TiO_2 content and treatment temperature: 1) unchanged glass; 2) darkening of color; 3) slight haze; 4) weak opalescence; 5) opalescence; 6) surface crystallization; 7) crystallization; 8) fine cracking; 9) large cracks; 10) color change.

Glass with 3% TiO_2 was found to behave similarly. In this case there was also a region of slight turbidity between the cracking and opaque glass regions.

The character of crystallization was sharply different in the case of glasses with 4-10% TiO_2. In all cases the region of outwardly unchanged glass was followed by a region of darkening. On further rise of temperature a region of cracking and slight opalescence was reached; this occurred at much lower temperatures than for glasses with lower TiO_2 contents. This was followed by a region of gradual decrease of opalescence, leading in some instances to total disappearance when observed visually. At and above 800°C the opalescence reappeared, increased rapidly, and at approximately 900°C the glass specimens became opaque and milky white throughout.

The behavior of glass with 11% TiO_2 was somewhat different. In this case the original glass was itself slightly opalescent. The cracking region was distinctly narrower.

The heat-treated glasses with various TiO_2 contents were investigated by a variety of methods.

When glass specimens without TiO_2 are heated in the 500-800°C range, they remain homogeneous and transparent. Visible crystallization begins at approximately 850°C. The results of electron microscope investigations are shown in Fig. 36. The micrographs show the presence of large blocks up to 10 μ in size at 1000°C. These blocks grow continuously with further increase of temperature to 1300°C. Electron micrographs of glass specimens with 1 and 2% TiO_2 are similar to the electron micrographs of the original glass without TiO_2.

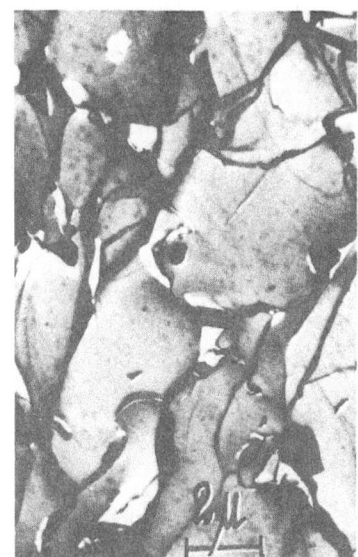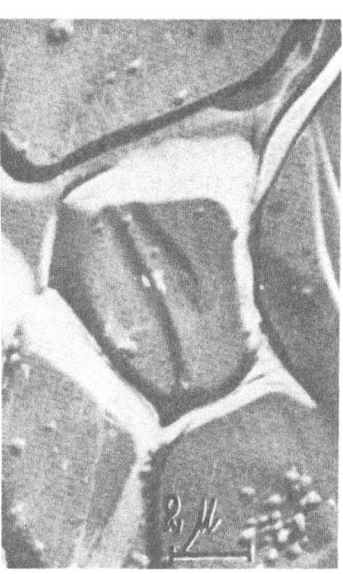

FIG. 36. Electron micrographs of glass specimens without TiO_2 after 24 hours of treatment in the gradient furnace: a) t ≈ 700°C; b) t ≈ 1000°C; c) t ≈ 1200°C.

Micrographs of glass containing 3% TiO_2 are shown in Fig. 37. No structural changes are observed in micrographs of specimens after heat treatment at 500-770°C. A microcrystalline phase separates out at 800°C (Fig. 37a). A structure consisting of large blocks is observed at higher temperatures; the blocks increase in size with increase of the temperature of heat treatment, reaching 15-17 μ at 1300°C (Fig. 37c). It may be noted that the blocks formed at various temperatures in glasses containing 1, 2, and 3% TiO_2 decrease in size with increase of TiO_2 content.

Photomicrographs of glass containing 5% TiO_2 (the original titanium-containing glass) are given in Figs. 23 and 26. As already noted (see Chapter V), at 680°C small formations appear and are collected in groups of various sizes. At 695°C these formations become considerably more numerous without increasing appreciably in size. Starting from 705°C, the structure becomes even more fine-grained (approximately 0.1 μ), and up to 820°C there are no noticeable changes in the numbers or size of the grains. The structure becomes very much coarser in the 820-1000°C range. Blocks 0.2-0.5 μ in size, consisting of fine formations, appear at 910°C. At 1000°C the structure of the specimen consists of large (0.4-0.6 μ) regions with noticeable faceting. A small amount of the microcrystalline phase is present. A certain amount of prismatic crystals of a third phase appears at the same time. At 1000-1300°C the crystals of all three types increase appreciably in size.

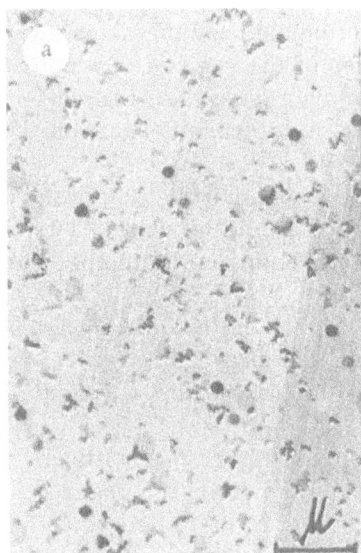 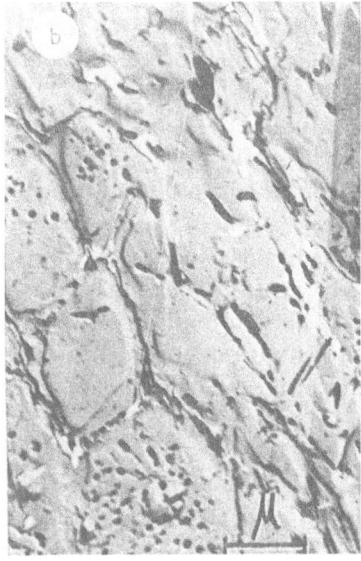 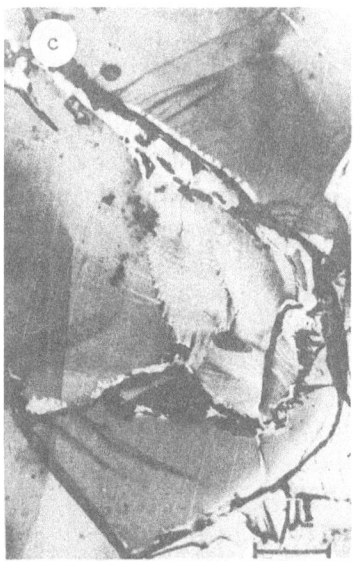

FIG. 37. Electron micrographs of glass specimens containing 3% TiO$_2$ after 24 hours of treatment in the gradient furnace: a) t ≈ 800°C; b) t ≈ 1000°C; c) t ≈ 1200°C.

The crystallization and recrystallization processes occurring in glass with 6% TiO$_2$ during heat treatment are approximately the same as in the glass with 5% TiO$_2$.

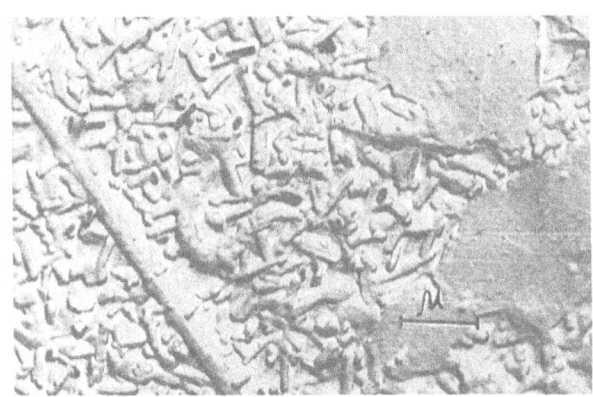

FIG. 38. Electron micrograph of glass specimen containing 8% TiO$_2$ after 24 hours of treatment in the gradient furnace. Large blocks and numerous prismatic crystals, t ≈ 1150°C.

The behavior of glass with 8% TiO$_2$ is similar. In addition, in this case flattened regions 0.5-10 μ in size appear on the general microcrystalline background at approximately 1000°C, and a considerable amount of prismatic crystals of various sizes, mainly approximately 0.8 μ, appears at approximately 1150°C in addition to the large blocks (Fig. 38).

Optical crystallographic investigation of glasses containing 0-3% TiO$_2$ showed that uniaxial crystals of negative optical sign separate out as the principal phase in all the glasses in the 850-950°C range. Examination of glasses subjected to heat treatment at 1000°C and higher temperatures showed that in this case the crystalline phase also consists of uniaxial crystals, but of positive optical sign. The crystal size decreases with increase of TiO$_2$ content, and increases with rise of treatment temperature. Sections of glasses crystallized at 1000-1200°C contained large crystal platelets of irregular shape but with distinct straight faces. The crystals often had fibrous structure. At temperatures close to the liquidus skeletal forms of crystal growth were observed. Under these conditions, fibers differing in refractive index from the main medium clearly could be seen growing through the fields of single crystals. After treatment at the liquidus boundary, the glass contained crystals in the form of tetragonal bipyramids surrounded by a crystalline border with a different refractive index. Similar crystallization of glasses close to spodumene in composition was observed by Hatch in a study of the system LiO$_2$ − Al$_2$O$_3$ − SiO$_2$ [1].

Determination of the optical characteristics of the crystalline phases formed in the glass without TiO$_2$ and with 1, 2, and 3% TiO$_2$ showed that the optical sign of the crystals of the main crystalline phase changes in a definite temperature region which depends on the TiO$_2$ content in the glass. The optical sign of the crystals changes from negative to positive on increase of temperature. This change occurs in the 970-1000°C range in the glass without TiO$_2$. The transition temperature for glasses with 1 and 2% TiO$_2$ as in the 920-950°C range. With 3% TiO$_2$ the lower limit becomes approximately 910°C.

Comparison of the optical-crystallographic and X-ray diffraction data for the same specimens made it possible to extend the determination of the boundaries of the regions in which the optical sign of the crystals changes to glasses with 4% TiO$_2$ and over from the X-ray diffraction data alone. Optical determinations could not be carried out in this case because of the very small amount of the crystalline phase formed.

It was found that introduction of 4% TiO$_2$ into the glass lowers the transition temperature range sharply (800-820°C). Glasses with 5 and 6% TiO$_2$ have the same transition range. Further increase of the TiO$_2$ content in the glass to 10% has no appreciable influence on the temperature region in which the optical sign of the crystals changes.

The results of optical and X-ray determinations of the ranges in which the optical sign of the crystalline phase changes in glasses with different TiO$_2$ contents are shown schematically in Fig. 39. For comparison, the temperature range of the transition in glass with the composition of spodumene is also shown.

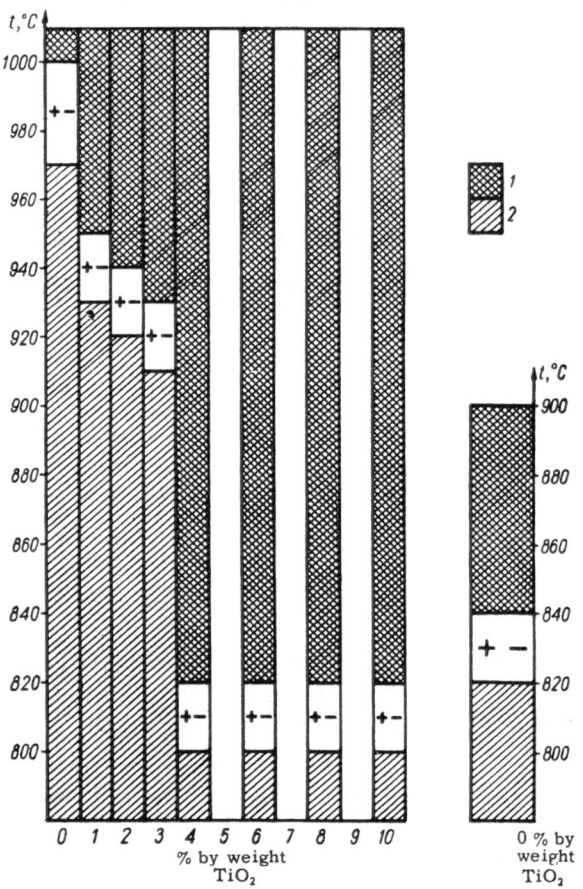

FIG. 39. Effect of the TiO$_2$ content in glass on the change of the optical sign of the crystals with change of temperature: 1) positive crystals; 2) negative crystals.

It is clear from the figure that with approximately 5% TiO$_2$ in the glass the change from negative to positive crystals occurs under the same temperature conditions in glass with the composition of spodumene.

Optical investigations show that the sharp change of sign of the main crystalline phase occurs in a relatively narrow temperature range (20-30°C), regardless of the TiO$_2$ content in the glass. In the transition range both negative and positive crystals can be distinctly recognized by the optical method, and their relative amounts alter progressively. Under certain temperature conditions, the contents of both can probably be the same.

The presence of two crystalline forms is recorded over a somewhat wider temperature range by careful X-ray diffraction analysis. Thus, mixtures of crystals in various proportions probably exist in a somewhat wider temperature range.

For determination of the composition of the prismatic crystals formed at temperatures above 1000°C in glasses with high TiO$_2$ contents, the original glass (with 5% TiO$_2$) was held in a furnace at 1000°C for 1, 13, and 25 days. Examination under the microscope showed that after 1 day of heat treatment the slightly opalescent glass contained numerous minute crystals with a high refractive index. After 13 days the number of these crystals increased considerably, and they became larger. The crystals were larger still after 25 days of heat treatment.

The crystalline phase under investigation was isolated by treatment of the glass with hydrofluoric acid. When the residue from glass which had been subjected to 25 days of heat treatment was examined under the microscope, an accumulation of fine prismatic crystals was observed. However, they were too small for optical crystallographic determinations.

For chemical determination of the composition of the crystalline phase, glass specimens containing 11% TiO$_2$ were used in the first instance; when examined under the microscope, they were found to contain similar small prismatic crystals of high refractive index. The crushed glass was treated twice with 20% HF solution on heating, until the glassy phase had been decomposed. After decantation and additional treatment with hot 15% HCl, the crystals were washed with water until neutral and were examined under the microscope. They were then fused with sodium carbonate, and the melt was dissolved in acid and analyzed. The crystals were found to consist of TiO_2 and Al_2O_3. The analytical results are given in Table 7.*

The discrepancies between the results of the individual experiments may be attributed to difficulties in separation and analysis of such a microcrystalline phase, but the results are close in order of magnitude. The crystals isolated from glasses containing 9, 7, and 3% TiO_2 also consisted of TiO_2 and Al_2O_3, but the relative proportions of the oxides were somewhat different, with a higher Al_2O_3 content (see Table 8).

The results in Table 8 are in agreement among themselves but differ from the data of Table 7. Thus, separation of the fine crystals and chemical determination of their composition showed that crystals the overall chemical composition of which can be represented as

TABLE 7
Chemical Composition of Crystals Formed During
Heat Treatment of Glass with 11% TiO_2 by Weight

| Sample No. | Oxide contents | | | | Amount of crystalline residue after leaching, % of total weight |
| | % by weight | | mole % | | |
	TiO_2	Al_2O_3	TiO_2	Al_2O_3	
1	25.4	74.6	30.2	69.8	4.3
2	33.0	67.0	38.8	61.2	6.5
3	37.0	63.0	42.6	57.4	6.4

$nAl_2O_3:mTiO_2$ separate out of glass containing 3-11% TiO_2, in a definite temperature range. The relative proportions of the oxides vary in accordance with the TiO_2 content of the glass.

The phase diagram (Fig. 40) of the system $Al_2O_3 - TiO_2$ shows that only one binary compound, $Al_2O_3 \cdot TiO_2$, exists in the system. Accordingly, the oxide ratios found may be attributed either to simultaneous separation of several crystalline phases, TiO_2, Al_2O_3, $Al_2O_3 \cdot TiO_2$, in different proportions, dependent on the TiO_2 content in the glass, or to formation of new aluminum titanates.

TABLE 8
Chemical Composition of Crystals Formed During
Heat Treatment of Glasses with Different TiO_2 Contents

| TiO_2 content in glass, % by weight | Sample No. | Oxide contents | | | | Amount of crystalline residue after leaching, % of total weight |
| | | % by weight | | mole % | | |
		TiO_2	Al_2O_3	TiO_2	Al_2O_3	
9	1	10.7	89.3	13.3	86.7	9.0
7	1	10.1	90.9	12.5	87.5	4.0
7	2	12.6	87.4	15.5	84.5	3.7
3	1	15.7	84.3	18.2	81.8	2.5
3	2	14.5	85.5	17.8	82.2	—

Two kinds of crystals—prismatic and laminar rhomboidal—can be clearly seen in electron micrographs of the crystals obtained from glass with 11% TiO_2 (Fig. 41).

*The analyses were performed by K. A. Yakovleva.

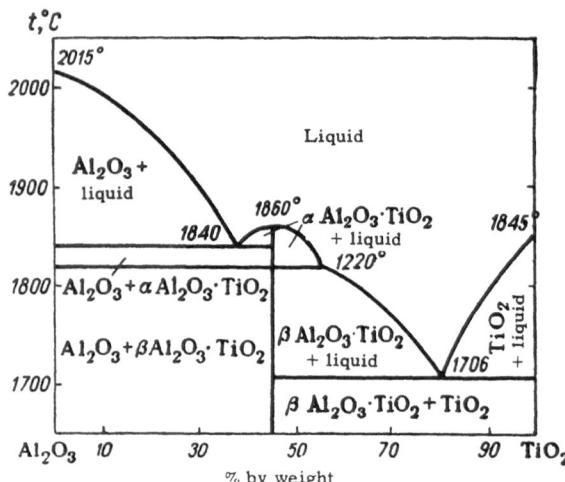

FIG. 40. Phase diagram of the system $Al_2O_3-TiO_2$.

FIG. 41. Electron micrographs of crystals isolated from glass with 11% TiO_2.

For a more complete study, the crystalline residue obtained by treatment of crystallized glass containing 11% TiO_2 with HF was investigated by chemical and Raman spectroscopy.

Spectrum analysis of the crystalline residue showed the presence of titanium, aluminum, and silicon and absence of lithium. Quantitative determinations of the individual elements were not performed. The presence of silicon in the residue may be attributed to the possible presence of a small amount of glassy phase or mullite.

In order to determine whether minute crystals (of the composition $nAl_2O_3 \cdot mTiO_2$) of the type detected in the region of visible crystallization are present in the temperature range in which transparent crystalline glass is obtained, the fully symmetric vibration bands in the Raman spectrum of the crystalline substance were compared with the spectrum of the transparent crystallized glass of the same composition with 5% TiO_2 (Fig. 42).* The spectrum of the transparent crystallized glass has a very strong line at 120 cm^{-2}. There is also a fairly distinct line at 620 cm^{-1}. The line at 120 cm^{-1} is not found in the region of the spectrum $nAl_2O_3 \cdot mTiO_2$ crystals adjacent to the exciting line. A band at 610 cm^{-1} can be seen in another region of the spectrum (recorded several times). These data show that the transparent crystallized glass does not contain the crystalline phase ($nAl_2O_3 \cdot mTiO_2$) [2].

The character of the Raman spectra confirms that the crystals are a titanium-containing phase, and shows that the structure of the phase resembles the rutile structure [3]. Since the investigations indicate with certainty that the phase contains TiO_2, it is described here as the titanium-containing phase until its composition and structure have been finally established.

Glasses containing from 0 to 11% TiO_2 were investigated by thermal analysis, and thermograms were recorded at temperatures up to 1100°C with an average heating rate of 30°/min.

*The determinations were performed by Ya.S. Bobovich.

TABLE 9

Characteristics of the Thermograms Recorded for the Glass Specimens at an Average Heating Rate of 30°/min

TiO_2 content, % by weight	First exothermic effect			Second exothermic effect		
	t_0, °C	Δt_{max}, °C	τ_{max}	t_0, °C	Δt_{max}, °C	τ_{max}
0	900	17	4 min 18 sec	—	—	—
1	880	14	4 ″ 06 ″	—	—	—
2	880	25	3 ″ 36 ″	—	—	—
3	880	24	3 ″ 12 ″	—	—	—
4—8	802	70	1 ″ 18 ″	996	8	1 min 18 sec
9	800	49	1 ″ 48 ″	1000	7	1 ″ 12 ″
10	765	23	1 ″ 06 ″	1010	4	3 ″ 06 ″

The thermograms for glasses with 0–3% TiO_2 were similar in appearance to the curve shown in Fig. 43a. The thermograms have one fairly flat and extended exothermic maximum.

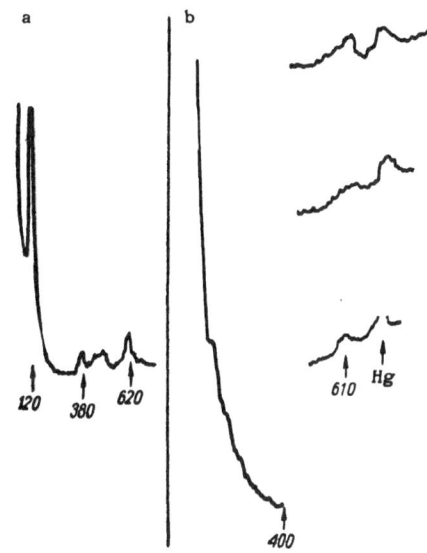

FIG. 42. Raman spectra: a) specimen of transparent crystallized glass with 5% TiO_2; b) crystalline product $nAl_2O_3 \cdot mTiO_2$.

The characteristics of the thermograms (t_0, Δt_{max}, and τ_{max}) are given in Table 9.

Since the heat effects for glasses containing from 4 to 8% TiO_2 are almost the same, average values of all three characteristics are given in the table.

It follows from the form of the thermograms (see Fig. 43a and b) and from the characteristics in Table 9 that only one phase crystallizes out in glasses with titanium contents of up to 3% inclusive. The glass grains merely become covered with a crystalline film; i.e., surface crystallization occurs. Therefore, relatively little heat is evolved and the exothermic effect is small and extended in time.

The appearance of two exothermic maxima at a TiO_2 content of 4% indicates that here two phases crystallize in the glass, crystallization becomes of the volume type, the quantity of heat evolved increases, and the process becomes considerably more rapid. The temperature at which the first exothermic effect begins (the crystallization region of the first phase) is lowered by 80–100°C. The second phase crystallizes at approximately 1000°C; i.e., it is more refractory than the first. The first exothermic effect begins to diminish at a TiO_2 content of 9%.

Blocks cast from glass containing 11% TiO_2 crystallized almost completely during annealing. Glass remained only in the outside layer. This layer was used for the thermal analysis. It was reasonable to suppose that the decrease of the heat effect is due to partial crystallization of the glass. Rapidly cooled specimens of this glass were free from visible crystallization. Rapidly cooled specimens of glasses with 10 and 11% TiO_2 gave the same first heat effect as glasses with 4–8% TiO_2. Consequently, in glasses containing 9% TiO_2 and over, crystallization may occur while the melt is cooling.

TABLE 10

Characteristics of the Glasses after Cooling

TiO_2 content, % by weight	Cooling				
	not held	held at 820°C	held at 740°C	held at 715°C	held at 700°C
3.0	Glass	Glass	Glass	Glass	Glass
5.0	Glass	Glass	Crystallized waxlike glass	Glass	Glass
7.0	Crystallized waxlike glass	Crystallized porcelainlike glass	Crystallized porcelainlike glass	Crystallized waxlike glass	Crystallized waxlike glass
9.0	Crystallized porcelainlike glass	The same	The same	–	–
11.0	The same	The same	The same	–	–

The thermogram in Fig. 43c shows that the second exothermic effect for glasses containing 10 and 11% TiO_2 has an additional plateau and becomes a doublet. This form of heat effect indicates that two phases with very similar crystallization temperatures separate out in this temperature region in glasses with high TiO_2 contents.

A series of experiments was carried out for elucidation of the part played by the TiO_2 catalyst and on the effect of the amount present on possible crystallization of the glass during cooling of the original melt. Glasses containing 3, 5, 7, and 11% TiO_2 were melted simultane-

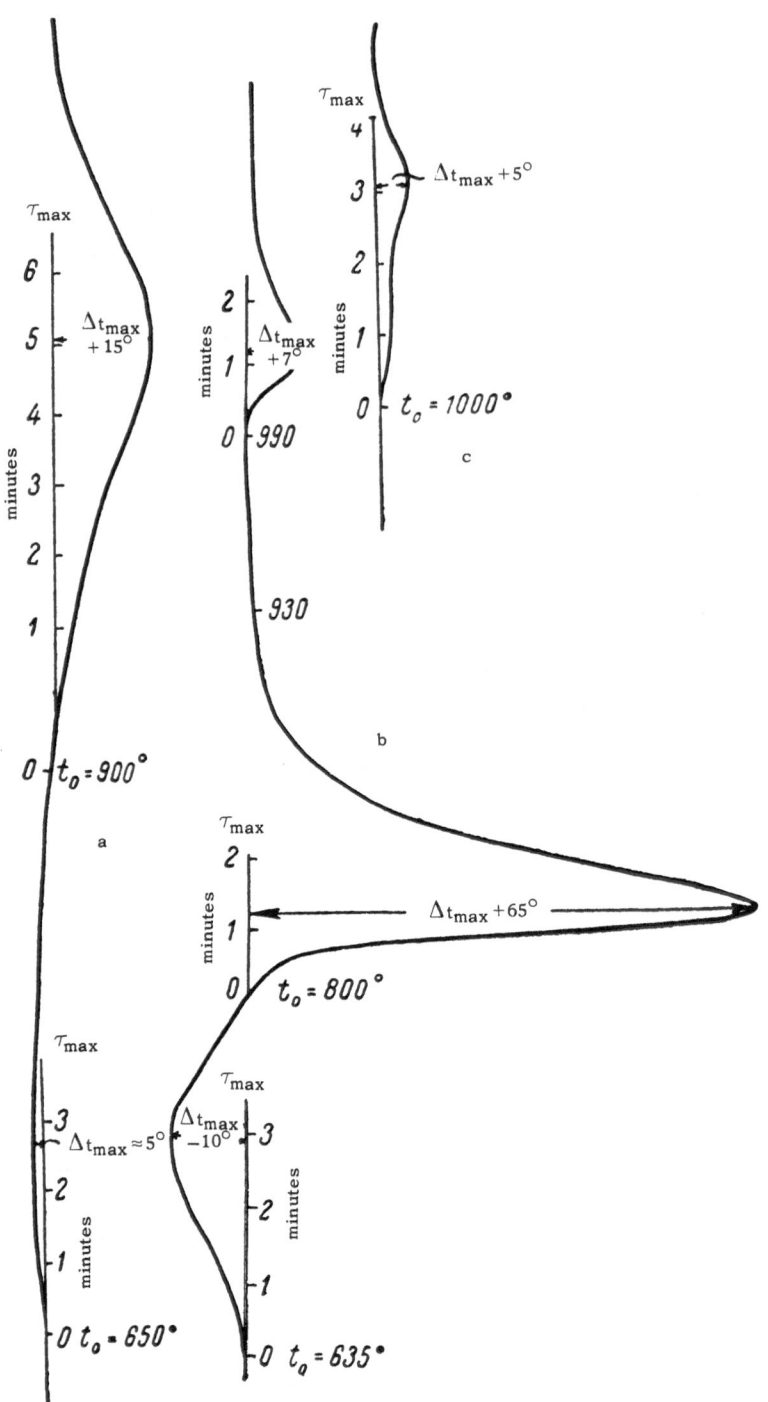

FIG. 43. Thermograms of glasses with different TiO_2 contents: a) 0-3%;
b) 4-8%; c) 10-11%.

ously in a furnace in 3-liter pots. At the end of the melting, at a temperature of 1480-1500°C
in the glass, the pots were transferred to an electric furnace preheated to 900°C and cooled according to definite schedules, which were checked with the aid of thermocouples immersed in
the glass melt. The cooling schedules for the different experiments are given in Table 10.

The crystallinity of the glasses was assessed visually when the pots cooled. The results
of this assessment are also given in Table 10. The table shows that the glass with 3% TiO_2 remains transparent under any of the cooling conditions studied. Glass with 5% TiO_2 crystallized
only when held at 740°C. Crystallization does not occur if this glass is held at temperatures

above or below 740°C. Glasses with 7, 9, and 11% TiO_2 crystallized to some extent under all the cooling conditions studied.

Crystallization occurs in a cooling glass only if the nucleation rate and rate of crystal growth are sufficiently high at a certain temperature. For this to occur, the rate curves for nucleation and for crystal growth must overlap in a certain temperature range. The temperature of the maximum crystallization rate probably corresponds to the point of intersection of these curves.

The glass with 3% TiO_2 does not crystallize under any of the cooling conditions chosen; this indicates that the above-mentioned curves do not intersect in this case.

The glass with 5% TiO_2 crystallizes only if held at approximately 740°C; therefore, the curves for this glass should intersect at about this temperature and overlap a narrow temperature range. Glasses with 7, 9, and 11% TiO_2 crystallize on cooling even if not held at a definite temperature, and the crystallization intensifies with increasing TiO_2 content. This means that the rate curves for nucleation and for crystal growth overlap to an increasing extent with increasing TiO_2 content; their point of intersection apparently remains the same as before, but the temperature region in which both processes occur at a high enough rate gradually widens with increase of TiO_2 content.

Under given crystallization conditions, dependent on which of the two rates is the higher, either numerous nuclei are formed in the glass but grow slowly, or there are fewer nuclei, of which the crystals grow to a larger size. This is reflected in the external appearance of the crystallized glass; in the former case it is waxy in appearance, and in the latter it is more like porcelain [4].

These investigations showed that the character of the crystallization process varies in accordance with the amount of titanium dioxide in the glass. Volume crystallization in the glass of the composition investigated occurs only when the titanium dioxide content is over 3%. At lower TiO_2 contents crystallization begins at the surface and gradually extends into the glass specimen. The tendency of the glass to crystallize increases with increasing TiO_2 content.

LITERATURE CITED

1. H. A. Hatch. Am. Mineralogist 28:471 (1943).
2. N. E. Kind. in: The Glassy State, Part 1, Izd. Akad. Nauk SSSR (1963), p. 105. [English translation: The Structure of Glass, Vol. 3, Consultants Bureau, New York (1964) p. 111.]
3. Ya. S. Bobovich and D. K. Arkhipenko. Opt. i Spektroskopiya 17(5) (1964).
4. E. V. Podushko and A. B. Kozlova. in: The Glassy State, Part 1, Izd. Akad. Nauk SSSR (1963), p. 74. [English translation: The Structure of Glass, Vol. 3, Consultants Bureau, New York (1964) p. 77.]

CHAPTER VII

Investigation of Crystalline Lithium Aluminosilicates

An investigator conducting phase analysis must have at his disposal the standard characteristics of the substances which are expected to separate out in the materials under investigation. In accordance with the phase diagram given in the literature for the system $Li_2O - Al_2O_3 - SiO_2$, the separation of β-eucryptite, β-spodumene, or of their solid solutions with silica, is to be expected in the materials being studied [1-3].

The first preliminary experiments confirmed that our original glass does not contain the low-temperature α forms of these compounds. The crystallization products in the glass resemble the high-temperature β forms of eucryptite and spodumene, although in a number of cases the agreement of their characteristics is not complete [4,5].

A more detailed study was carried out of the crystallization products of glasses in the system $Li_2O - Al_2O_3 - SiO_2$, corresponding in composition to definite chemical compounds (eucryptite and spodumene) or close to them.

Glass specimens from different meltings, both in the original state and crystallized, were investigated by a variety of methods.

X-ray diffraction studies were carried out on specimens of glass with the composition of eucryptite, crystallized in a boat in a gradient furnace at 600-1000°C, and also on a number of specimens crystallized at temperatures of the order of 900-1000°C, as listed below.

1. A glass specimen with the composition of eucryptite, crystallized by slow cooling in the temperature region near the liquidus (the structure of the specimen was microcrystalline).
2. Compact crystalline aggregates with well-defined cleavage, grown during cooling of a glass of eucryptite composition melted in a 3-liter quartz pot.
3. A crystallized specimen of eucryptite glass, obtained from the Art Glass Factory as a standard eucryptite specimen (microcrystalline structure).

The X-ray diffraction diagrams obtained for the glass specimens crystallized in boats are given in Fig. 44. They are all in good agreement; this indicates that the same crystalline phase separates out over the whole temperature region investigated. The amount of this phase increases with rise of temperature. The X-ray diffraction diagrams of the other specimens are all similar to those shown in Fig. 44 and all coincide with the X-ray diffraction diagrams of β-eucryptite given in the literature [2-6].

Crystal-optical analysis of specimens crystallized in boats at 500-1000°C showed that the crystalline phase which forms in them consists of lamellar crystals with negative optical sign and refractive indices $n_g \approx 1.524$ and $n'_p \approx 1.518$. No changes of refractive index or optical sign of the crystals were observed in the heat-treatment temperature range studied. This phase also coincides in its optical characteristics with literature data for β-eucryptite (see Table 3).

Crystal-optical analysis of individual specimens crystallized at temperatures of the order of 900-1000°C also confirms that the characteristics of the crystals formed are in agreement with literature data for β-eucryptite [2,7].

Examination of sections prepared from the compact crystalline aggregates grown during cooling of eucryptite glass in a 3-liter pot showed that they consist of crystals oriented in one

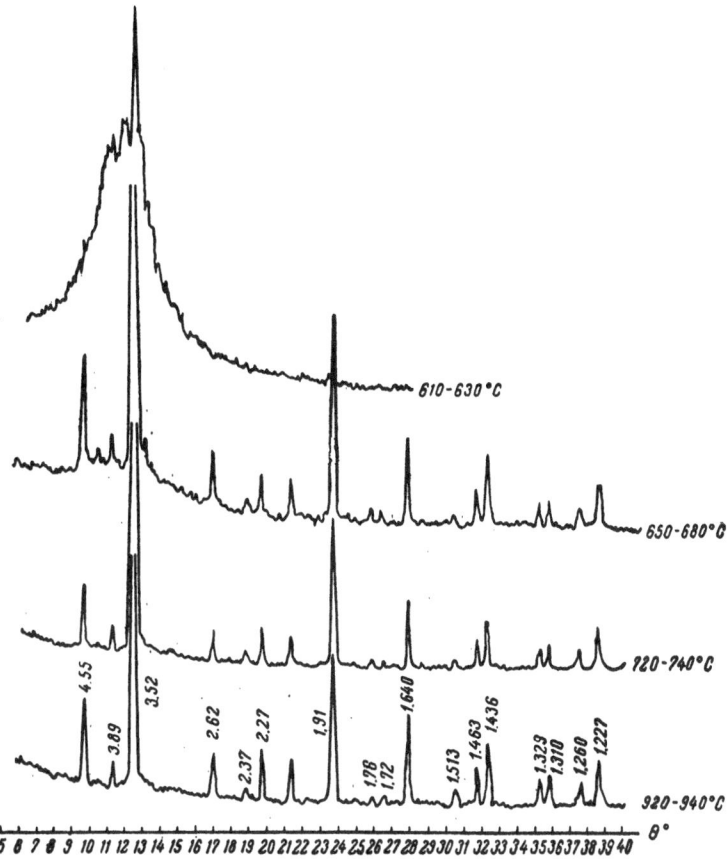

FIG. 44. X-ray diffraction diagrams of glass of eucryptite composition after heat treatment in the gradient furnace in the 600–1000°C temperature region for 24 hours.

plane. These aggregates also contain cavities and small foreign crystalline inclusions, the total content of which is 10-15%. The presence of foreign inclusions is an indication of the inadequate purity of the specimens, although a second phase was not recorded by X-ray diffraction analysis.

The infrared reflection spectra of the specimens listed above were also recorded (Fig. 45).

In the case of the above-mentioned crystalline aggregates, reflection spectra were obtained from two mutually perpendicular planes (curves 5 and 6).

Moreover, for comparison of the transmission spectrum of β-eucryptite reported in the literature [8] with our reflection spectra of various crystallized specimens of eucryptite glass, we determined the transmission spectra of a specimen crystallized near the liquidus (curve 4) and of a single crystal of β-eucryptite, grown in the Institute of Silicate Chemistry* (curve 2).

The positions of the maxima in the reflection spectra given in Fig. 45 (curves 1,3,5) coincide. The two transmission spectra (curves 2 and 4) also agree with each other and with the transmission spectrum of β-eucryptite [8]. Since the reflection (curve 1) and transmission (curve 4) spectra refer to the same specimen and agree with the reflection spectra of curves 3, 5, and 6 and the transmission spectrum of curve 2, they are evidently all in mutual agreement and agree with the transmission spectrum of β-eucryptite [8]. It follows that the phases separating out in our crystallized specimens are a form of β-eucryptite.

However, investigations of the infrared spectra of the surface layers of eucryptite glass specimens crystallized at temperatures below 760°C in a number of cases yield a spectrum differing in character from those shown in Fig. 46 (curves 4 and 5: spectra with two bands in

*Specimen kindly provided by V. A. Ioffe.

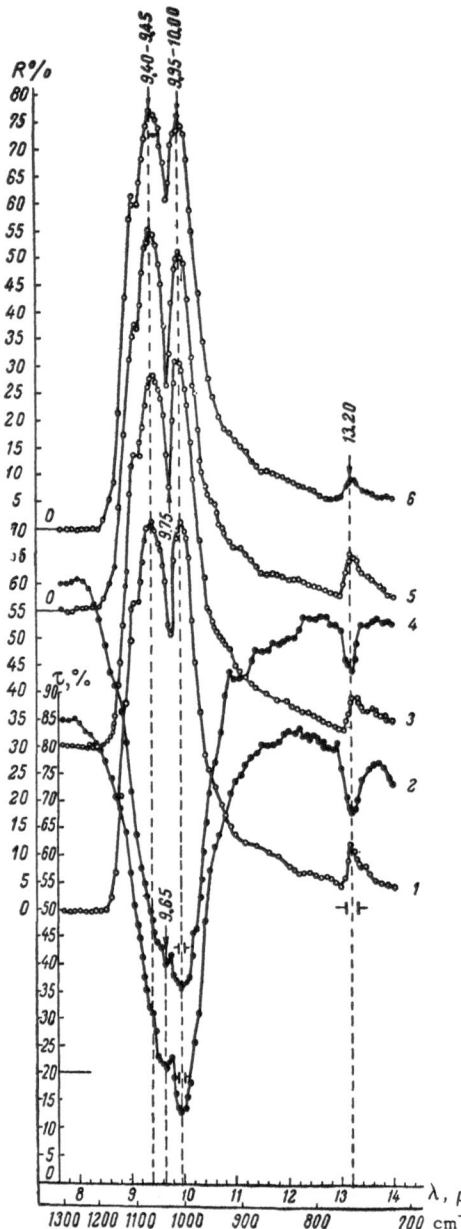

FIG. 45. Infrared reflection and transmission spectra of eucryptite glasses from different sources, crystallized at high temperatures. Reflection spectra: 1) specimen crystallized near the liquidus; 3) specimen from the Art Glass Factory; 5,6) large crystals from a 3-liter pot. The specimens were cut in mutually perpendicular planes. Transmission spectra: 2) crystal from the Institute of Silicate Chemistry; 4) specimen crystallized near the liquidus.

the region of 9–10μ). This spectrum has one reflection band, the maximum of which at 9.8μ corresponds to the minimum between the maxima in the spectrum with two bands (curves 4 and 5). For brevity, we will refer to these as one- and two-band spectra.

One-band spectra were obtained in some instances also for specimens of eucryptite glass with addition of 5% TiO_2 (curve 2) and for glass close to eucryptite in composition, with addition of 5% TiO_2 (curve 3), crystallized at temperatures below 760°C.

It should be noted that the crystalline phase giving a spectrum with one band is fairly stable in certain cases. For example, a glass specimen close to eucryptite in composition, containing 5% TiO_2, which remained transparent after crystallization at 750°C, was heated again at approximately 900°C for about 3 hours. It remained transparent, although a crystalline haze appeared on its surface.

Figure 46 (curve 6) shows that the principal band at 9.8μ in the spectrum of this haze begins to split into two bands, the maxima of which coincide with the maxima of the two-band spectra. The spectrum of the deeper layers remain of the one-band type. When the specimen is heated further at 1000 and 1100°C for 6 hours, it becomes opaque. However, even in this case the spectrum of the inner layers remains of the one-band type; this shows that the main mass of the specimen did not crystallize.

The relative intensities of the bands vary in spectra with two maxima, obtained from crystallized specimens of eucryptite glass (Fig. 47) either from the same or from different meltings. Most of the spectra also have an intermediate band. This suggests that in the case of a spectrum with two bands each band corresponds to a separate crystalline phase, i.e., that such specimens contain two main phases simultaneously. A glass specimen giving a spectrum with one band contains only one main phase. All three phases differ in structure. It is natural to suppose that the phase giving a simple spectrum with one band is a compound, and in this case is one of the eucryptite phases.

Comparison of the X-ray diffraction diagrams of specimens giving one- and two-band infrared reflection spectra showed that the diffraction lines were shifted somewhat in relation to each other (Fig. 48). For more precise clarification of this shift, an X-ray diffraction diagram was obtained for a mixture containing 50% of each of the specimens; this diagram is shown in Fig. 48 (curve III). It was found that far from all the lines undergo the shift, which does not conform to the relationships given in the literature for solid solutions with a eucryptite-like structure.

It must be emphasized that in the X-ray diffraction diagrams of specimens giving one-band infrared spectra some of the lines were always shifted in the same direction and to the same extent. The constant shift of the same individual lines suggests that here we have two

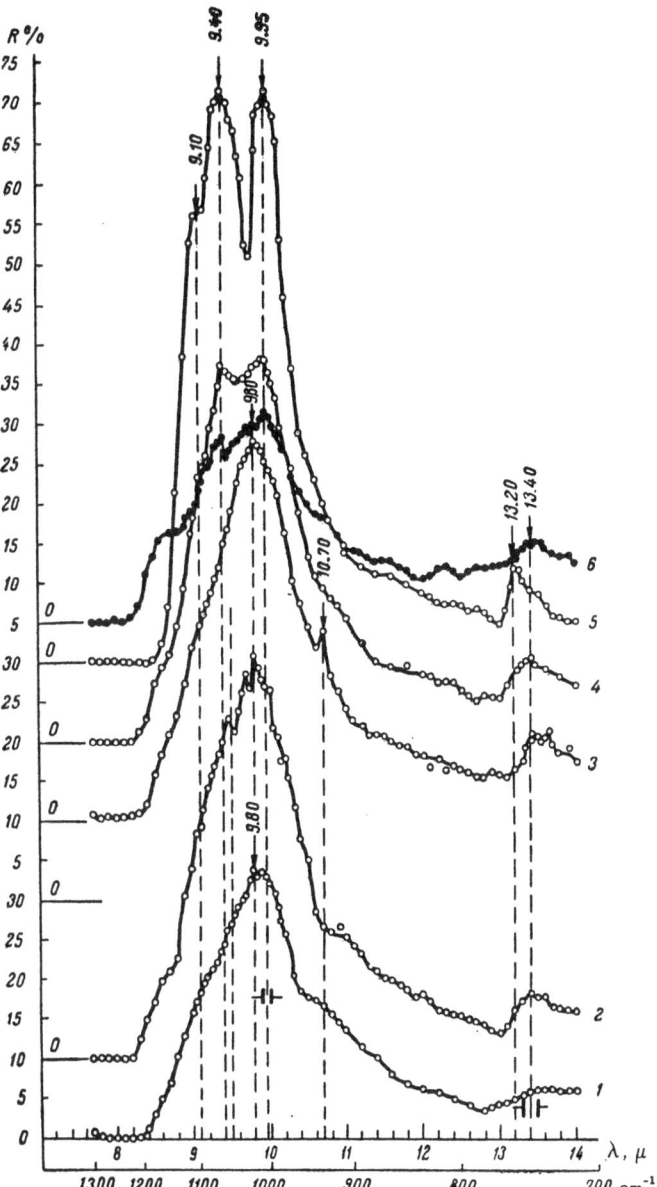

FIG. 46. Infrared reflection spectra of glass specimens of or near the composition of eucryptite, crystallized under different conditions: 1,6) surface layers; 2,3,4,5) inner layers; 1) t = 755°C, τ = 40 min; 2,3) t = 750°C, τ = 24 h; 4) t = 1410–1350°C, τ = 6 h; 5) specimen crystallized near the liquidus; 6) t = 750°C, τ = 24 h; and t = 900°C, τ = 3 h.

different forms of eucryptite rather than lattice deformations; the latter can cause only random line shifts, although of the same magnitude as were found in the present instance.

The X-ray diffraction diagram of the glass specimens giving infrared spectra with two bands is in better aggreement with the published diagrams of β-eucryptite than the X-ray diffraction diagrams of specimens giving infrared spectra with a single band.

The observed shift of individual lines in the X-ray diffraction diagrams of specimens giving one-band infrared spectra may be attributed to a change in one of the lattice parameters while the lattice type remains the same. Calculation showed that the presumed β'-eucryptite form has a unit cell shorter by approximately 3% along the C axis than the β-eucryptite cell.

Differential thermograms were recorded in the 400-1100°C range for two meltings of eucryptite glass; one is shown in Fig. 49.

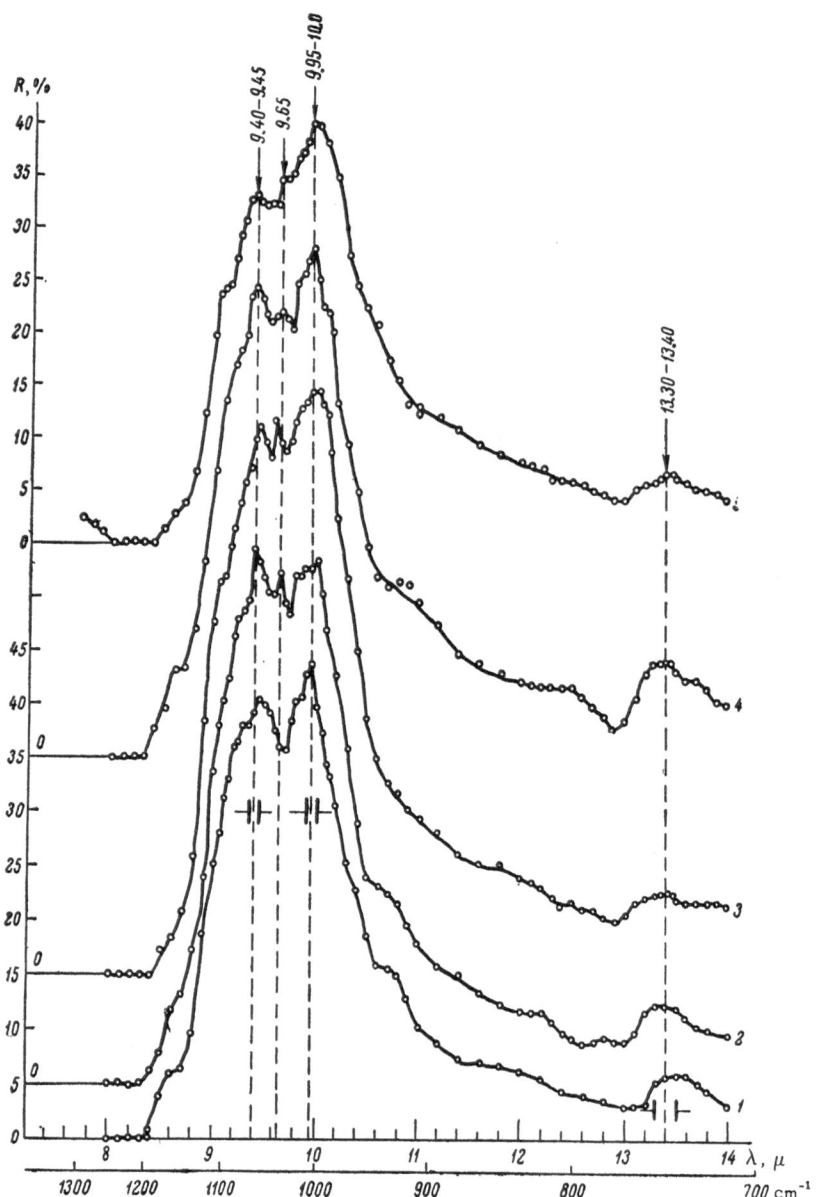

FIG. 47. Infrared reflection spectra of crystallized glass specimens of eucryptite composition from different meltings: 1,2) surface and inner portions of the crystallized glass (without added TiO_2); for curve 1, t = 900°C, τ = 9 h; for curve 2, t = 1000°C, τ = 24 h; 3,4) glass crystallized during casting (with addition of 5% TiO_2); for curve 3 the cut is parallel and for curve 4 it is perpendicular to the crystal fibers; 5) crystallized glass (with addition of 5% TiO_2), t = 1000°C, τ = 24 h.

It follows from Fig. 49 that the endothermic effect usually observed in glass in the annealing range changes to an exothermic effect due to evolution of heat during crystallization of the first phase.

The presence of a single narrow exothermic peak with a large temperature deviation indicates that one main crystalline phase separates out in glass with the composition of eucryptite at temperatures up to 1100°C. If any additional phase does crystallize out, the amount is so small that the heat evolved during differential thermal analysis is not detected; i.e., if any other phases are present, they are there as minor impurities.

Thus, it may be concluded from the above data that one stable crystalline phase separates out in glass of eucryptite composition during crystallization in the 600-1100°C range. In some

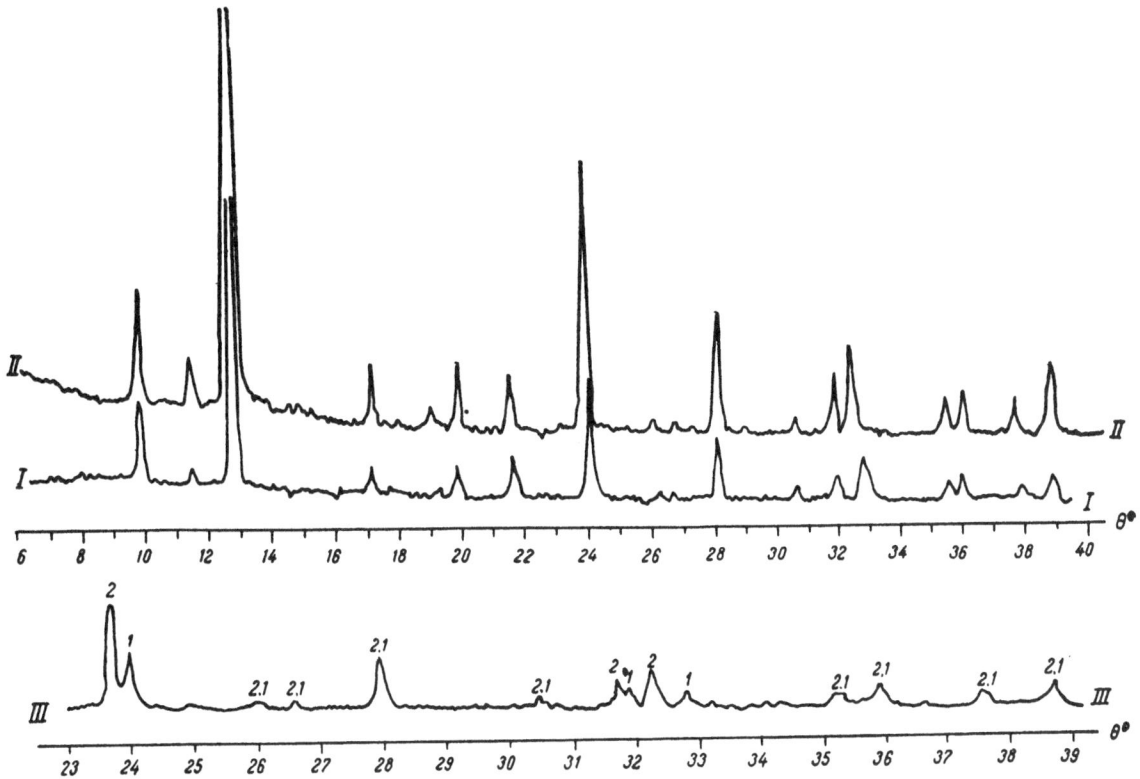

FIG. 48. X-ray diffraction diagrams of crystallized specimens of eucryptite glass giving one-band (I) and two-band (II) infrared reflection spectra, and of a mixture of these specimens (III). The numerals 1 and 2 on X-ray diagram III indicate maxima corresponding to the maxima of diagrams I and II.

cases another form of eucryptite may separate out, differing in its infrared spectrum and X-ray diffraction diagram from the known α and β forms of eucryptite.

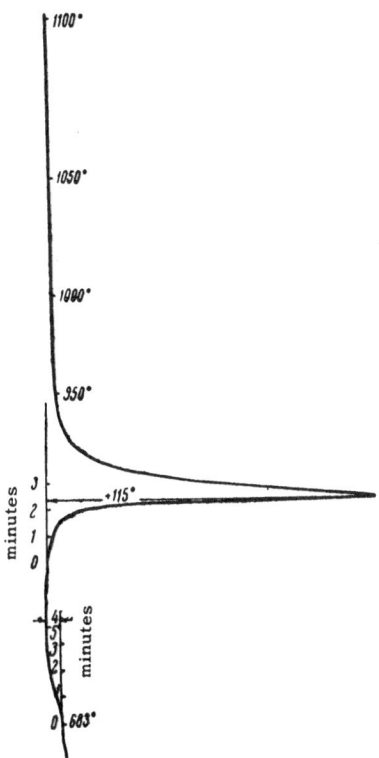

FIG. 49. Thermogram of glass with the composition of eucryptite.

The phase which usually separates out in the entire temperature range indicated during crystallization of glass of eucryptite composition coincides in its X-ray diffraction diagram, optical characteristics, and infrared spectrum with the phase known in the literature as β-eucryptite. Although the two-band structure of the infrared spectrum of β-eucryptite suggests that in this temperature range it is not a compound but is actually a mixture of two phases, until this hypothesis has been finally confirmed we will accept the generally adopted view and assume that we are dealing with a single phase, referred to as β-eucryptite. The second phase, detected in individual experiments, can logically be named β'-eucryptite, and we adopt this name for that phase.

A crystal of native α-spodumene mineral was used in the first instance for obtaining the standard characteristics of β-spodumene. Its X-ray diffraction diagram is shown in Fig. 50.*

The α-spodumene was converted into the high-temperature β-form by heat treatment at a series of temperatures from 900 to 1150°C for different times.

*Crystals of α-spodumene from different origins differ in their impurity contents; therefore the X-ray diffraction diagram of our crystal deviates somewhat from the X-ray data in the literature for native α-spodumene crystals and for specimens synthesized by the hydrothermal method [2].

X-ray diffraction diagrams, also given in Fig. 50, were obtained from the α-spodumene crystal at different stages of heating and at the end of the process. Comparison of these diagrams shows that transition of the low-temperature α form of spodumene into the high-temperature β form begins at about 900°C. Heat treatment in the region of 950°C results mainly in the formation of an intermediate phase, the X-ray diffraction diagram of which differs from the diagram for β-spodumene given in the literature [8-10].

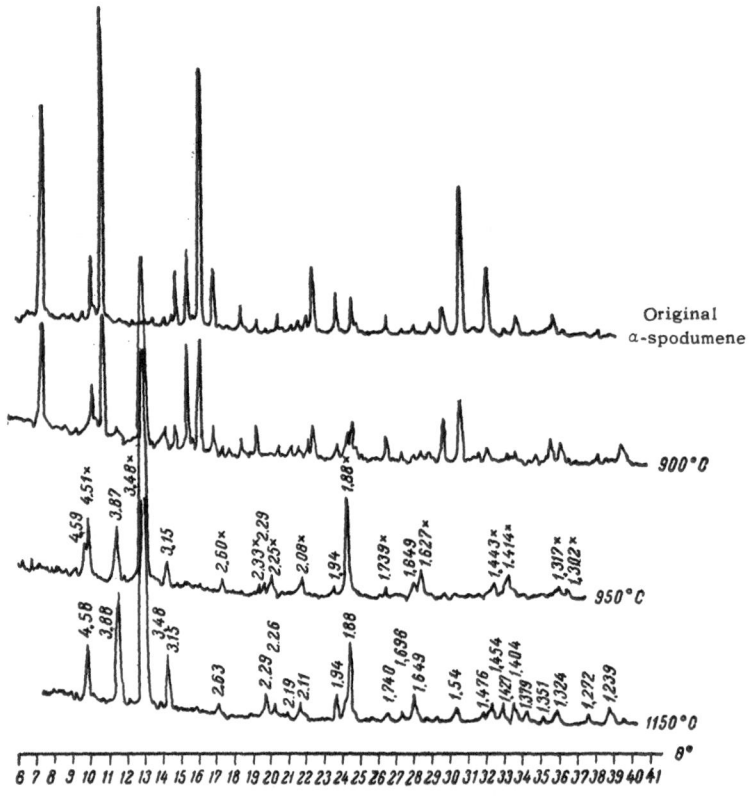

FIG. 50. X-ray diffraction diagrams of α-spodumene crystals, in the original state and after heat treatment at various temperatures.

The principal phase appearing at approximately 1150°C gives an X-ray diagram which coincides satisfactorily with the diagram for β-spodumene known from the literature, and it can therefore be adopted as the standard.

The quantitative proportions of the two phases alter in the range between 900 and 1150°C: the amount of intermediate phase decreases and the amount of the phase corresponding to β-spodumene increases with rise of temperature.

In addition to a native α-spodumene crystal, crystallization products of glass with the composition of spodumene were also investigated. Specimens of this glass were crystallized in a boat in the gradient furnace in the 500-1000°C range, and also under constant-temperature conditions at 720, 740, 760, 780, 800, 820, 840, and 860°C for 24 hours.

Crystal-optical analysis of the specimens crystallized in the boat and at constant temperatures showed that lamellar crystals with $n_g \approx 1.518$ and low birefringence separate out over the entire temperature range of 500-1000°C. (Table 3 shows that n_g values of this order are characteristic of β-spodumene and its solid solutions.) Up to 830°C the crystals are optically negative. A crystalline optically negative phase with the above value of n_g is mentioned in the literature as negative β-spodumene [2,11]. More recently it has been regarded as a solid solution of eucryptite with silica [3]. However, for brevity we provisionally retain the name of the negative β-spodumene for this phase.

Crystal-optical investigations of specimens crystallized at the constant temperatures stated above showed that negative β-spodumene persists as the principal phase up to 830°C. At

830-840°C the optical sign of the crystals changes to positive. Above 840°C negative β-spodumene changes into the known high-temperature form of β-spodumene. The two phases coexist in the 830-840°C range; the amount of the first phase falls rapidly while the amount of the second increases at the same rate with rise of temperature.

Figure 51 gives the X-ray diffraction diagrams and crystallization temperatures of the crystallized specimens. The diagrams 4 and 5, for crystals formed at temperatures above 840°C, differ from diagrams 1 and 2 for crystals formed below 830°C. The two former agree with each other, with the diagram for 1150°C in Fig. 50 which we adopted as the standard, and with the X-ray diagram of β-spodumene given in the literature [2].

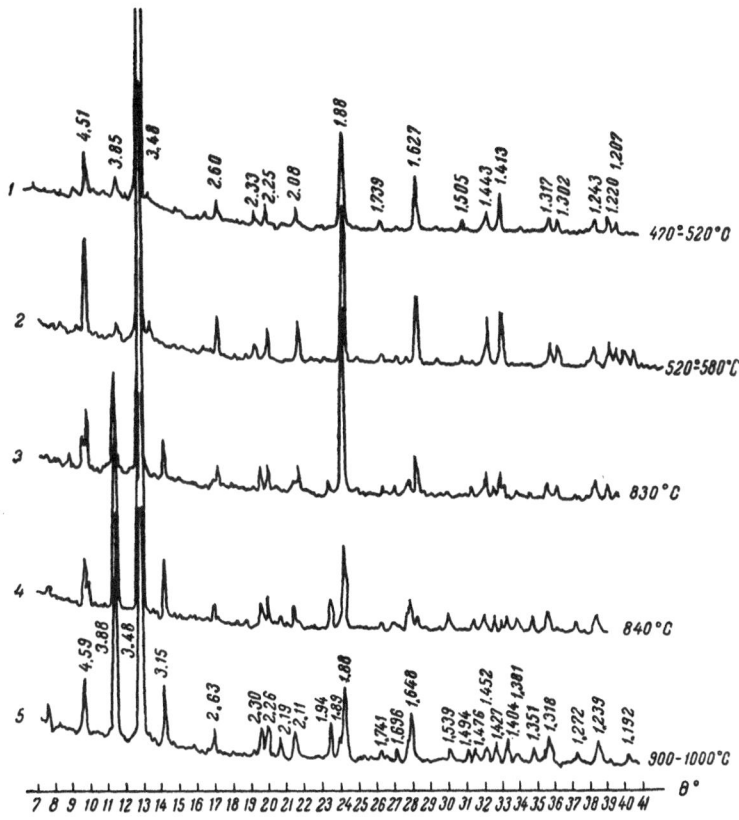

FIG. 51. X-ray diffraction diagrams of spodumene glass crystallized at various temperatures for 24 hours.

Diagrams 1 and 2 in Fig. 51 agree with each other in the number and positions of the lines, and coincide fairly closely with the X-ray diagram of negative β-spodumene given in the literature [2]. At the same time, these diagrams resemble that of β-eucryptite (Fig. 44). They differ from the latter by a shift of the diffraction lines toward larger angles, i.e., they exhibit the characteristics of diagrams for solid solutions.

Thus, our data indicating the closeness of negative β-spodumene to β-eucryptite agree with Roy's conclusion that negative β-spodumene belongs to the β-eucryptite series ("O" series) of solid solutions [3]. Accordingly, negative β-spodumene should be regarded as a member of the eucryptite series of solid solutions, with the composition of spodumene or close to it, and having a eucryptite-like structure.

The following fact is worthy of attention. Comparison of the X-ray diffraction diagrams of a crystal of α-spodumene heated at 950°C (Fig. 50) and of negative β-spodumene (Fig. 51) shows that the intermediate phase formed during the conversion of the α-spodumene crystal into β-spodumene is none other than negative β-spodumene (the lines of negative β-spodumene are indicated by crosses on the X-ray diagram for 950°C in Fig. 50).

Infrared reflection spectra were obtained from glass specimens of spodumene composition without TiO_2 and with addition of 5% TiO_2, crystallized in boats. The most characteristic of these spectra are given in Fig. 52. Investigation of the infrared spectra showed that specimens crystallized below 810°C give spectra of the form of curve 1. These spectra have two sharp maxima, at $9.30-9.35\mu$ and $9.75-9.80\mu$, in the $9-11\mu$ region.

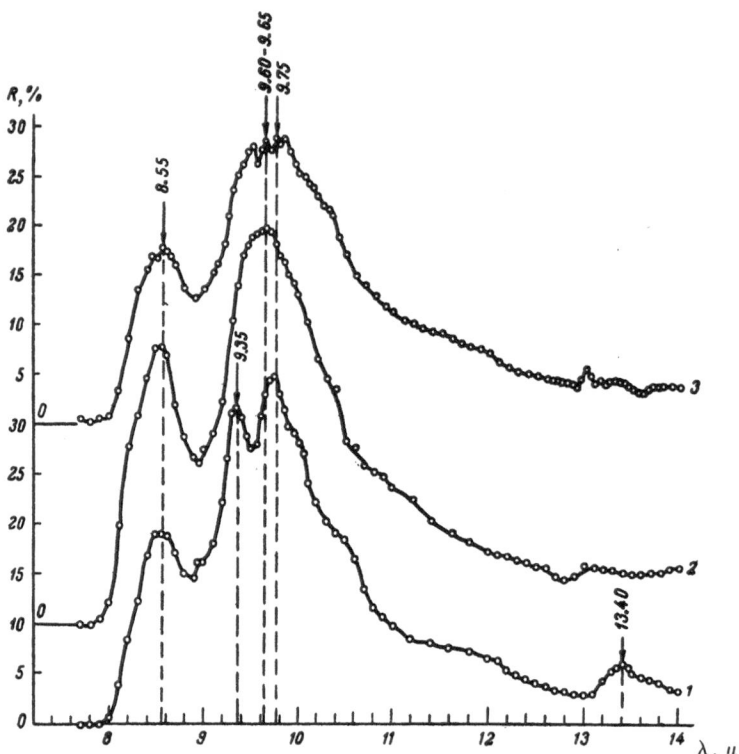

FIG. 52. Infrared reflection spectra of glass specimens of spodumene composition, crystallized at different temperatures: 1) below 810°C; 2) above 850°C; 3) at 1100°C; 3 min.

These maxima gradually level out with rise of temperature above 810°C; above 850°C they are replaced in the reflection spectrum (Fig. 52, curve 2) by a broad diffuse band with a reflection maximum at 9.6μ, i.e., where the spectrum for lower temperatures has a reflection minimum. Curve 2 resembles in general outline the spectrum given in the literature [12] for β-spodumene, with a broad band at $9-10\mu$ and a short-wave band at 8.6μ.

Comparison of curve 1 with the spectra of numerous glasses, both with the composition of spodumene and differing somewhat from it, suggests that the crystallization products giving spectra similar to curve 1 are mixtures of silicates of different chemical composition.

When the temperature is raised a solid-phase reaction probably occurs, giving a new phase which is characterized by spectrum 2 (Fig. 52). A series of intermediate compounds is formed in the course of this reaction, and as a result a mixture of crystals with spectrum 2 or 3 appears. This is supported by the following facts: 1) the very broad and diffuse outline of the band, which may be due to superposition of spectra of crystals with bands close together; 2) the variable position of the maximum of this diffuse band, between 9.5 and 9.8μ; 3) the presence, in a number of cases, of a well-defined structure in this band—additional peaks of variable positions.

Moreover, it may be concluded from our investigation of the crystallization products of the system $Li_2O - SiO_2$ that they often contain one of the forms of silica (but not quartz, tridymite, or cristobalite) which has a selective reflection maximum at $8.55-8.60$ μ. The short-wave band at 8.50μ which is always found in the infrared spectra of the crystallization products of spodumene glass coincides in the position of its maximum with the above-mentioned silica band.

This suggests the presence of the same form of silica as an admixture. This suggestion is also supported by the fact that the position of this band is unaltered regardless of whether there is a doublet or one broad band in the $9-10\mu$ region.

Since these views on the complex structure of the two spodumene phases are based only on infrared spectroscopic data and have not yet been confirmed by other methods of investigation, before this problem has been finally solved we will consider each phase as crystals of one type, retaining the names of negative and positive β-spodumene for them.

Transition between the two phases occurs in a relatively narrow temperature range (830-840°C).

Only negative β-spodumene crystallizes below 830°C, with $n_g \approx 1.518$, a negative optical sign, and the X-ray diffraction diagram and two-band reflection spectrum given in Figs. 51 and 52 (curves 1).

Only positive β-spodumene crystallizes above 840°C. It has the value $n_g \approx 1.518$ and a positive optical sign; its X-ray diagram is given in Fig. 51. Its infrared reflection spectrum has a broad and diffuse principal maximum (Fig. 52, curve 2).

Glasses of spodumene composition were also investigated by thermal analysis; the thermogram obtained resembles in appearance the thermogram of eucryptite glass shown in Fig. 49. The thermogram has only one narrow exothermic peak in the 400-1100°C range, with the following characteristics: start of the exothermic effect, $t_0 = 830$°C; maximum temperature deviation, $\Delta t_{max} = 95$°C; duration of the effect, $\tau_{max} = 1$ min 40 sec.

The absence of any other exothermic peaks shows that only one crystalline phase separates out in spodumene glass in the temperature range studied. It follows from the foregoing that this phase has the composition of spodumene, but it cannot be determined from the thermogram which of the high-temperature forms separates out during the thermal analysis. In any event, it may be said that the exothermic peak observed at about 830°C for spodumene glass corresponds to crystallization of a principal phase with the composition of spodumene or close to it.

The following conclusions can be drawn from the foregoing.

1. In the 600-1100°C temperature range the crystallization product of glass with the composition of eucryptite is β-eucryptite. The optical constants, X-ray diffraction diagram, and two-band infrared reflection spectrum of this phase are given above. In some instances another phase may form, described by us as β'-eucryptite, having a one-band infrared reflection spectrum (Fig. 46, curve 1) and the X-ray diffraction diagram shown in Fig. 48.

2. The crystallization products of glass with the composition of spodumene include two phases: negative β-spodumene (a member of the eucryptite series of solid solutions), formed at temperatures below 830°C, and positive β-spodumene, formed at temperatures above 840°C. Their optical characteristics, X-ray diffraction diagrams, and infrared reflection spectra are given above.

The characteristics of these phases given in the present chapter have been adopted by us as the standard characteristics and will be used for analysis of the crystallization products of the glass under investigation.

LITERATURE CITED

1. H. A. Hatch. Am. Mineralogist 28:471 (1943).
2. R. Roy, D. M. Roy, and E. F. Osborn. J. Am. Ceram. Soc. 33:152 (1950).
3. R. Roy, Z. Krist. 14:185 (1959).
4. A. G. Alekseev and L. A. Fedorova. in: The Glassy State, Part 1, Izd. Akad. Nauk SSSR (1963), p. 84. [English translation: The Structure of Glass, Vol. 3, Consultants Bureau, New York (1964) p. 90.]

5. V. A. Florinskaya, E. V. Podushko, I. N. Gonek, and E. F. Cherneva. in: The Glassy State, Part 1, Izd. Akad. Nauk SSSR (1963), p. 90. [English translation: The Structure of Glass, Vol. 3, Consultants Bureau, New York (1964) p. 96.]

6. H. G. Winkler. Acta Cryst. 1:27 (1948).

7. D. S. Belyankin, V. V. Lapin, and N. A. Toropov. Physicochemical Principles of Silicate Technology, Promstroiizdat (1954).

8. V. A. Kolesova. Izv. Akad. Nauk SSSR, Ser. Khim. Nauk, No. 1 (1963).

9. B. Skinner and H. J. Evans. Am. J. Sci. 25(8A):312 (1960).

10. V. N. Plyushchev, Yu. P. Simanov, and I. V. Shakhno. Dokl. Akad. Nauk SSSR 125:334 (1959).

11. E. Heinglein. Fortschr. Mineral. 34:40 (1956).

12. M. K. Murthy and E. M. Kirby. J. Am. Ceram. Soc. 45(7) (1962).

CHAPTER VIII

Crystallization of Lithium Aluminosilicate Glass

In this chapter we summarize all the experimental material presented in the book and present an analysis of the structural changes taking place in the glass under investigation during heat treatment and of the composition of the crystalline phases formed.

No noticeable structural changes were detected in glass subjected to heat treatment at 600°C for 24 hours. We may therefore assume that it is in the same state as the original glass, and that its properties can be compared with those of specimens after heat treatment at higher temperatures. However, this does not mean that structural processes cannot occur in this temperature region. Experiments show that when glass is held at 600°C for 200 hours or more, its physical state changes; but these changes are slight, occur very slowly, and are disregarded when we examine the kinetics of the process.

The first distinct changes are found in specimens after heat treatment in the precrystallization range (630-700°C). Starting from 630°C, the optical density, density, refractive index, and leachability change and the specimens become yellowish brown. At the same time the color of the glass containing Ni^{2+} changes from yellow to greenish.

Starting from 630°C, as observed in the electron micrographs, these regions are replaced by small formations, approximately 0.1μ in size, which at first form separate groups; when the temperature is raised to 695-700°C they rapidly become more numerous (without noticeable change of size) and fill the whole volume of the glass.

However, the X-ray diffraction diagrams, infrared absorption spectra, the magnitude of the heat effect, and linear thermal expansion remain the same in this temperature range of heat treatment as those of the original glass; this indicates absence of a crystallization effect. Therefore it may be assumed on the basis of the electron micrographs that separation into at least two phases occurs in this temperature region [1-3].

The causes of separation of the glass into components differing in composition are as yet not clear, but it may be definitely stated that in the case of the glass under investigation volume crystallization is associated with preceding phase separation and that the decisive part is played in this by the amount of TiO_2 introduced into the glass. The results of investigations of glasses with different TiO_2 contents (see Chapter VI) show that phase separation regions do not appear in glass without TiO_2 or in glasses containing up to 3% TiO_2. These regions first appear in glass with 4-5% TiO_2, when volume crystallization also occurs.

A necessary condition for production of transparent crystalline glass materials is microseparation with regions approximately 0.1μ in size, preceding crystallization. In studies of glasses in the system $Li_2O - Al_2O_3 - SiO_2$ we found that such fine separation occurs when the glass composition lies to the right of the $SiO_2 - Li_2O \cdot Al_2O_3$ line (Fig. 22), i.e., in the field where aluminum appears in sixfold coordination.

It follows from visual observations that a yellow-brown color appears in the glass when its TiO_2 content is raised to 4%; the color is appreciably intensified on further increase of the TiO_2 content. It is due to the presence of TiO_2 and Fe^{2+} in the glass [4,5]. Similar intensification of the color of this glass occurs as the result of heat treatment at temperatures above 600°C, although the amounts of TiO_2 and Fe^{2+} do not increase. In our opinion, the explanation of this contradiction is that phase separation occurred in the glass with TiO$_2$ and Fe^{2+} entering one of the separated components, increasing their concentration in that component and deepening its color. The color of the second component becomes correspondingly weaker. However, as was shown earlier [6], the deepening of the yellowish brown color of the glass is not proportional to the TiO$_2$ and Fe^{2+} contents; the color deepens much more rapidly. Therefore, the greater the amount of these additives passing into the corresponding separating component during heat treatment of the glass, the more the total color of glass specimens in which separation has occurred should deepen. As will be shown later, the most probable supposition is that TiO$_2$ and Fe^{2+} enter the component containing Al$_2$O$_3$ and form a structure with the elements in sixfold coordination.

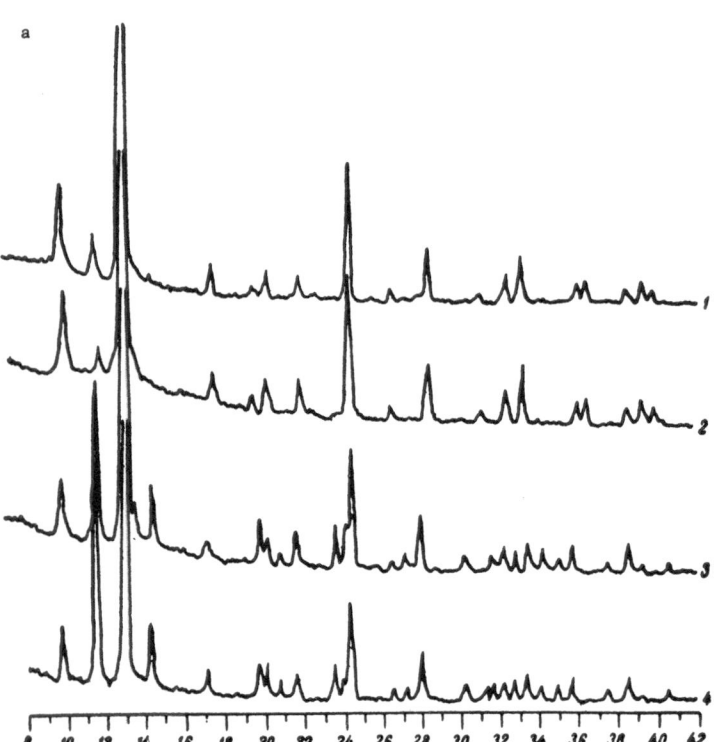

FIG. 53. X-ray diffraction diagrams (a) and infrared reflection spectra (b) of specimens of the original glass and of glasses with the composition of spodumene, crystallized at different temperatures: 1) spodumene glass crystallized at 820°C (negative β-spodumene); 2) original glass crystallized at a temperature below 820°C (transparent); 3) original glass crystallized above 820°C (opaque); 4) spodumene glass crystallized above 820°C (high-temperature spodumene).

The change of the yellow color of glass containing Ni^{2+} to green as the result of heat treatment indicates that Ni^{2+} passes completely into sixfold coordination in this temperature region.

The spectral curves of glass colored with Ni^{2+} indicate that nickel is present in the original glass mainly in sixfold coordination, with only traces in fourfold coordination. After heat treatment at 630°C the traces of nickel in fourfold coordination vanish entirely. A distinct minimum appears on the spectral curve in the 560 mμ region, and the glass becomes green. As the coordination of Ni^{2+} in glass of the given composition is associated with the coordination of aluminum, this change in the color of the glass is in good agreement with the presence of aluminum in sixfold coordination in the corresponding separating component.

Electron micrographs show that the individual small formations which appear at 680 and 695°C are similar in size to the formations observed at 705-720°C. From the appearance of the latter in the electron micrographs they cannot be said to be crystals.

However, it is quite evident from the results of X-ray structural analysis, analysis of the infrared spectra, thermal analysis, and changes in the nature of the linear thermal expansion that above 700°C the glass crystallizes. Since the crystals formed are comparable in size to the regions of phase separation (approximately 0.1μ) and since the crystallization extends throughout the volume, it may be assumed that above 700°C each individual small region of separation crystallizes, forming crystals of the same composition as the original separated phase. Under such conditions crystallization proceeds easily and does not require diffusion.

Comparison of the X-ray diagrams and infrared reflection spectra of the separated crystalline phase with the corresponding characteristics of the crystallization products of glasses with the compositions of eucryptite and spodumene shows that the principal crystalline phase formed is negative β-spodumene (see Fig. 53, curves 1 and 2), so that it is of the composition of spodumene or close to it. Since the crystals of this phase are formed from the regions of the principal separating phase, the latter must also be of the composition of spodumene, or close to it. This conclusion is confirmed, first, by the lowering of the region in which the optical sign of the crystallization products of the glass changes with increase of the TiO_2 content to 4-5% (which is accompanied by the appearance of phase separation) to the temperatures at which the optical sign of the crystals changes in crystallized spodumene glass, and, second, by the lowering of the heat effect of crystallization of the first phase to approximately 820°C, i.e., the temperature at which the corresponding heat effects begins in spodumene glass, on introduction of 4-5% TiO_2 into the glass.

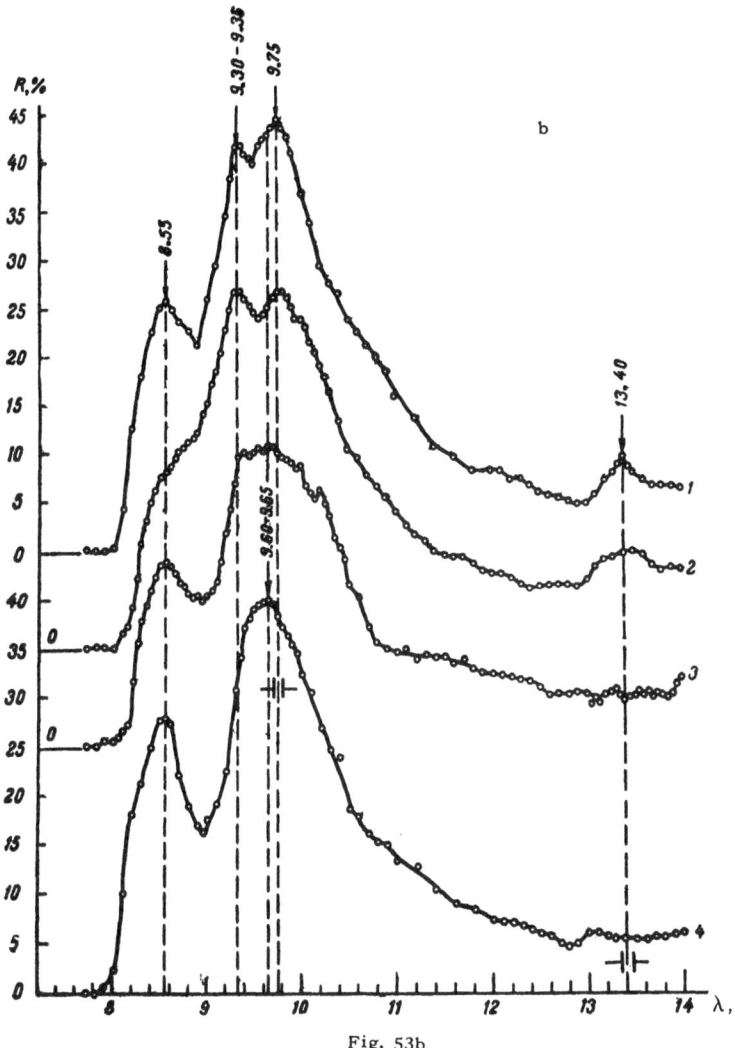

Fig. 53b

Thus, it may be concluded with adequate justification from an analysis of a number of observed facts that the glass separates into phases, and that the main separating phase consists of glass close to spodumene in composition.

It is logical to suppose that separation of these phases in the precrystallization range is preceded by formation of microheterogeneous regions of the corresponding composition while the glass is still in melt form, these regions being "frozen" when the glass is cooled [3,7-9]. In this sense the period of formation of the microheterogeneous regions may be described as the preseparation period.

During heat treatment in the precrystallization range, the titanium-containing regions, because of the high surface tension of this phase, apparently separate out before the principal spodumene phase. The formation of submicroseparation regions containing TiO_2 and Al_2O_3, uniformly distributed throughout the glass (not observed, because of their small size, in the electron micrographs), apparently assists the main process—microdisperse separation of the spodumene composition.

The phase containing TiO_2 and Al_2O_3 crystallizes at temperatures above 1000°C; this is shown by a series of experimental data. When the glass under investigation crystallizes in this temperature region, the appearance of a new crystalline phase can be detected by optical crystallography and electron microscopy. Chemical analysis of the crystals formed demonstrated the presence of Al_2O_3 and TiO_2, while investigation of their Raman spectra confirmed the presence of TiO_2 in the crystals. Crystallization of this phase is also confirmed by the appearance of a second exothermic maximum at about 1010°C on the thermogram of the glass.

Therefore the phase containing Al_2O_3 and TiO_2, which crystallizes at a higher temperature than the phase with the composition of spodumene, should evidently be in the amorphous state in the precrystallization range.

The nature of the processes occurring in the 600-700°C temperature range can be deduced on the basis of the above suggestion that the glass separates in the precrystallization range into components one of which is close to spodumene in composition and the other consisting of Al_2O_3 and TiO_2.

It was shown earlier (Chapter V) that throughout the transparent crystallization range (700-820°C) only the crystalline phase which was detected during crystallization in the 700-720°C range separates out of the original titanium glass. The size of the crystals formed also remains without noticeable change throughout this range. It is only when the glass is treated at 790°C that, as Fig. 23,9 shows, a certain orientation appears in the arrangement of the particles.

It follows from Fig. 34 that the physical properties of the glass change most sharply at 700-720°C. This also indicates that considerable structural changes occur in this temperature range.

The effect of preliminary heat treatment on the elongation of the glass is of considerable interest (Fig. 31). After heat treatment in the temperature region below 700°C the elongation of the glass during subsequent heating of the specimens to 400°C is the same as for the original glass, but it changes sharply after heat treatment of the specimens in the 700-720°C range. This fact indicates that the change is due to separation of the main crystalline phase (negative β-spodumene).

The only explanation of the change in the nature of the elongation can be that the crystals formed have a high negative coefficient of expansion and that the amount of crystalline phase gradually increases in the 700-720°C range. Of the crystalline aluminosilicates formed in the system $Li_2O - Al_2O_3 - SiO_2$ only β-eucryptite has a sufficiently large negative coefficient of expansion ($-70 \cdot 10^{-7}$) to account for the observed change in the character of the elongation of the crystallized specimens. The negative coefficient of expansion of β-spodumene (less than $0.5 \cdot 10^{-7}$) is too small for this. Therefore, by its value of the coefficient of expansion negative β-spodumene should be closer to β-eucryptite than to β-spodumene. This conclusion agrees with the results of X-ray structural analysis, indicating that negative β-spodumene has a structure of the eucryptite type.

Since an experimentally determined value of the coefficient of expansion of negative β-spodumene is not available, it is not possible to calculate the amount of the principal phase separating out. However, the fact that during crystallization in the 700-720°C range the coefficient of expansion alters from $+60 \cdot 10^{-7}$ to $-4 \cdot 10^{-7}$, with the assumption that the coefficient of expansion of negative β-spodumene is of the same order of magnitude as that of β-eucryptite, indicates that the amount of crystalline phase formed must be at least 50% of the total mass of the glass.

Glasses crystallized in the 720-760°C range exhibit some degree of further contraction; this indicates that crystallization of the principal phase continues to 760°C.

This conclusion is in good agreement with the calculated changes of the amount of crystalline phase in the glass in relation to the temperature of heat treatment, based on X-ray diffraction data.

The amount of crystalline phase was estimated by comparison of the X-ray diffraction patterns of the specimens with those of standards. The diffraction line observed at the Bragg angle θ_m characterizes the phase m in the given mixture. In the general case, the intensity of this line can be represented by the expression

$$i_m = C_m q_m K(\theta_m) Y_0$$

where C_m is the volume concentration of the given phase; q_m is a factor which depends on the nature of the phase and on the line chosen; $K(\theta_m)$ is the absorption factor; and Y_0 is the intensity of the primary beam.

If determination of i_m for the specimen is immediately followed by determination of the intensity i'_m of the same line for a standard specimen with a known concentration C'_m of the crystalline phase, we can write

$$\frac{i_m}{i'_m} = \frac{C_m}{C'_m} \cdot \frac{K(\theta_m)}{K'(\theta_m)}$$

It was assumed for the results of X-ray diffraction studies and from data obtained by other methods that when the original glass (with 5% TiO_2) crystallizes, the crystals which separate are mainly close in chemical composition to the original glass. In this case $K(\theta_m) \approx K'(\theta_m)$ and the pure phase can be used as the standard ($C'_m = 1$). Therefore, in our case $C_m = i_m/i'_m \cdot 100\%$, and the contents of crystalline phase in the specimens can be determined as the ratio of the most intense maxima in the X-ray diagrams of the crystallized original glass and the standard specimen. The crystalline compounds which were presumed to be present in the glass were chosen as the standards. Since these compounds were not available to us in the pure form, they were prepared by crystallization of glass with the composition of spodumene, and by heat treatment of low-temperature α-spodumene at 950 and 1100°C. The identity of these standards with the crystallization products of the specimens was checked by X-ray diffraction; the absence of amorphous phase was confirmed microscopically.

The calculated values were used for plotting the amount of principal crystalline phase as a function of the temperature of heat treatment (Fig. 54). The course of the curve shows that the main mass (up to 55%) of the crystalline phase separates out before 720°C is reached; at about 750°C the amount of crystalline phase reaches approximately 60%, and then remains almost unchanged with further rise of temperature.

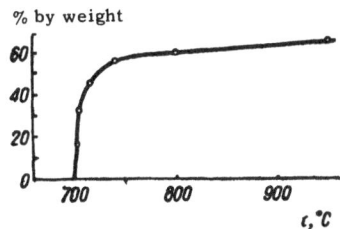

FIG. 54. Variation of the amount of crystalline phase in the original glass with the temperature of heat treatment.

The ratio of the intensities of the maxima in the infrared reflection spectra gives a somewhat lower amount of crystalline phase. In the case of minerals and well-crystallized glasses the reflection coefficient at the band maximum is of the order of 80-90%. For the original glass with 5% TiO_2 it is approximately 26%, and for transparent and opaque crystallized specimens it is only 34-36%. Therefore, according to the infrared reflection spectra, the amount of crystalline phase in the transparent and opaque crystalline glass material may be approximately 35%, which is considerably less than the value found from X-ray diffraction data. Both calculations involved somewhat crude approximations; these account for the discrepancy.

After heat treatment in the 760-820°C range the contraction of the specimens during the first stage of heating gradually decreases and the subsequent elongation increases. The explanation for this course of behavior upon heating is that during the preliminary heat treatment of the glass at temperatures above 760°C, a phase having the same course of expansion as negative β-spodumene is formed. The amount of this phase increases somewhat with rise of temperature to 820°C.

The cracking of the specimens in the crystallization range at 700-720°C, described in Chapter V, can also be attributed to the large negative coefficient of expansion of the crystals of the principal phase.

The situation may be represented as follows. At first, during crystallization, individual crystals with a considerable negative coefficient of expansion separate out in the main mass of glass with a large negative coefficient of expansion. This results in such large local stresses at the glass-crystal boundary that the specimen cracks on cooling. Subsequently, at temperatures approaching approximately 720°C, the cracking ceases; this indicates considerable decrease or total disappearance of the microstresses. At these temperatures nearly all the possible amount of crystals of the principal phase separates out. The coefficient of expansion of the remaining glass, containing Al_2O_3 and TiO_2, should differ from that of the first phase. The fact that the cracking stops indicates that the coefficient of expansion of the residual glass is appreciably less than that of the original glass.

When the treatment temperature is raised to 820°C and higher, i.e., in the range of opaque crystallization, the glass becomes very turbid and then opaque. Radical changes in the glass structure in this range can be detected by various methods.

Electron micrographs show some coarsening of the crystalline phase at 820°C. The nature of the X-ray diffraction diagrams and infrared spectra also changes. The X-ray diagrams and infrared spectra of the glass crystallized at 820°C (Fig. 53a and b, curves 3) coincide with those of high-temperature β-spodumene which separates out when glass with the composition of spodumene crystallizes at temperatures above 820°C (Fig. 53a and b, curves 4). This shows the presence of high-temperature β-spodumene in the glass crystallized at 820°C.

It was shown in Chapter VI that the optical sign of the crystals in the glass under investigation changes from negative to positive at 800-820°C. The elongation curves of glass specimens crystallized at 800-820°C are similar in character to the curve for high-temperature β-spodumene [10].

All the above facts show that high-temperature β-spodumene appears in the glass during crystallization in the 800-820°C range as the result of recrystallization from negative β-spodumene.

Electron micrographs of specimens crystallized at 900°C show the presence of large blocks (up to $0.4\,\mu$) consisting of fine crystals at 900°C, and the appearance of large crystals with distinct faceting at 1000°C. The principal crystalline phase (high-temperature β-spodumene), which in this case consists of large lamellar crystals up to $1\,\mu$ in size, is accompanied in the photographs by a distinct microcrystalline phase, concentrated mainly in the centers of the large crystals. These crystals are apparently the residues of the microcrystalline phase of the negative β-spodumene which existed at lower temperatures. Finally, the micrographs also clearly show crystals of prismatic form, which are the titanium-containing product of crystallization of the second phase formed during the phase separation.

When the crystallization temperature is raised to 900°C the course of expansion of the glass on heating again alters noticeably (Fig. 31). Contraction no longer occurs at the start of the heating, and the specimen begins to expand at once. The coefficient of expansion in the range up to 400°C is $17 \cdot 10^{-7}$, which is less by a factor of about 3.5 than that of the original glass. A specimen crystallized at 1000°C gave the same elongation curve as a specimen crystallized at 900°C.

The fact that the linear expansion of specimens crystallized at 900 and 1000°C is the same shows that no new phases are formed and the relative amounts of the phases present are unchanged in this temperature range. These conclusions are in good agreement with the other data for the same glass specimens.

* * *

This investigation demonstrated the value of applying a series of different physicochemical methods to the study of a single material. This broad investigation yielded many interesting results which extend our understanding of the volume crystallization process. The results obtained in a study of a single glass are of more general character and may be extended to crystallization of other glasses in the given system.

However, despite the use of a variety of methods of investigation, some facts could not be explained. It is clear that a sufficiently complete picture of all the processes taking place in glass during heat treatment over a wide range of temperatures was difficult to obtain by a study of a glass of one composition.

Therefore, further investigations are being carried out on a large number of compositions in the system $LiO_2 - Al_2O_3 - SiO_2$, catalyzed by TiO_2, in the region of glass formation. Special attention is to be devoted to compositions with excess Al_2O_3 in relation to the join $LiO_2 \cdot Al_2O_3 - SiO_2$, which are of great interest for the development of a theory of catalyzed crystallization. The results will be presented in the second part of this book.

LITERATURE CITED

1. R. Roy. J. Am. Ceram. Soc. 43(12):670 (1960).
2. W. Vogel. Z. Chem. 3(7):271 (1963).
3. V. N. Filipovich. in: The Glassy State, Part 1, Izd. Akad. Nauk SSSR (1963), p. 9. [English translation: The Structure of Glass, Vol. 3, Consultants Bureau, New York (1964) p. 9.]
4. V. V. Vargin. Dokl. Akad. Nauk SSSR 103(1):105 (1955).
5. A. Dietzel and R. Boncke. Glastechn. Ber. 19:217 (1941).

6. V. V. Vargin. Production of Colored Glass, Gizlegprom (1940).

7. I. L. Krupatkin. Zh. Neorgan. Khim. 1(6):1210 (1956).

8. A. I. Avgustinik. in collection: The Glassy State, Izd. Akad. Nauk SSSR (1960), p. 115. [English translation: The Structure of Glass, Vol. 2, Consultants Bureau, New York (1960) p. 95.]

9. I. I. Kitaigorodskii, E. M. Rabinovich, and V. I. Shelyubskii. Steklo i Keram., No. 12:1 (1963).

10. I. M. Buzhinskii, E. I. Sabaeva, and A. N. Khomyakov. in: The Glassy State, Part 1, Izd. Akad. Nauk SSSR (1963), p. 127. [English translation: The Structure of Glass, Vol. 3, Consultants Bureau, New York (1964) p. 133.]

ADDITIONAL LITERATURE

E. V. Zhukovskii and D. I. Portugalov. Steklo i Keram., No. 5:41-43 (1958).

G. L. Mikhnevich. Kolloidn. Zh. 21(1) (1959).

K. T. Bondarev and V. A. Minakov. Steklo i Keram., No. 12:22 (1960).

K. T. Bondarev and R. Ya. Borodai. Steklo i Keram., No. 10:1-4 (1960).

F. K. Aleinikov and R. B. Slizhis. Dokl. Akad. Nauk SSSR 141(3):674 (1961).

I. I. Kitaigorodskii and P. I. Litvinov. Steklo. Byul. GIS, No. 3:1-4 (1961).

I. I. Kitaigorodskii and P. I. Litvinov. Steklo. Byul. GIS, No. 1:3 (1961).

G. Bliznakov. in collection: The Growth of Crystals, Vol. III, Izd. Akad Nauk SSSR (1961). [English translation: Consultants Bureau, New York (1962).]

E. S. Sorkin. Optiko-Mekhan. Prom., No. 10:33 (1962).

K. T. Bondareva and V. A. Minakov. Optiko-Mekhan. Prom., No. 9:26 (1962).

A. I. Korelova and O. S. Alekseeva. Optiko-Mekhan. Prom., No. 9:32 (1962).

N. N. Sirota. in collection: Crystallization and Phase Transitions, Izd. Akad. Nauk BSSR (1962), pp. 62-106.

V. G. Chistoserdov and I. A. Soboleva. Optiko-Mekhan. Prom., No. 9:35 (1962).

N. V. Dubovitskaya, E. É. Zasimchuk, L. N. Larikov, and Yu. N. Petrov. Ukr. Fiz. Zh. 8(10): 1134 (1962).

I. I. Kitaigorodskii, P. I. Litvinov, and L. S. Zevin. Steklo. Byul. GIS, No. 1:1 (1962).

F. Ya. Galakhov. Izv. Akad. Nauk SSSR, Otd. Khim. Nauk, No. 5:743 (1962).

Steklo i Keram., No. 12:3 (1962).

I. M. Vaisfel'd, A. A. Gorbachev, and L. M. Yusim. Dokl. Akad. Nauk SSSR 152(4) (1963).

Ya. S. Bobovich and G. T. Petrovskii. Zh. Strukt. Khim. 4(5):765 (1963).

G. M. Bartenev, A. I. Denishev, and A. I. Kolbasnikova. Steklo, Inform. Material GIS, No. 2 (1963).

Ya. S. Bobovich. Opt. i Spektroskopiya 14(5):647 (1963).

I. I. Kitaigorodskii. Zh. Vses. Khim. Obshchestva im D. I. Mendeleeva 8(2):192 (1963).

Steklo i Keram., No. 2:45 (1963).

L. G. Vaibursh and E. A. Fainberg. Steklo i Keram., No. 1:46 (1963).

E. A. Porai-Koshits. Optiko-Mekhan. Prom., No. 10:7 (1963).

A. G. Vlasov and T. E. Chebotareva. Optiko-Mekhan. Prom., No. 10:12 (1963).

S. D. Brown. J. Am. Ceram. Soc. 43(2):116-117 (1960).

B. Hinz. Silikat Tech., No. 3:119-122 (1959).

F. F. Fluss. Glass Instr. Techn. 2(1):14-16 (1958).

B. Semba. Szklo i Ceram. 12(9) (1961).

S. N. Lungu, D. Popesku-Has, and I. Teodoresku. Rev. Phys. 2(1):73 (1961).

W. Klemm and H. Welkmann. Glastech. Ber. 34(3):152-159 (1961).

M. K. Murthy and E. M. Kirby. J. Am. Ceram. Soc. 45(7):324 (1962).

B. Semba. Szklo i Ceram. 13(11):321 (1962).

J. Robredo. Verres Refractaires 16(5):273 (1962).

S. M. Ohlberg and D. W. Stricker. J. Am. Ceram Soc. 45(4):170 (1962).

J. E. Rindon. J. Am. Ceram. Soc. 45(1):7-12 (1962).

S. D. Stookey and K. D. Maurer. Proc. Ceram. Soc. 2 (1962).

K. A. Eppeler. J. Am. Ceram. Soc. 46(2):97 (1963).

M. D. Beals and J. H. Strimpl. Glass Ind. 44(9) (1963).